THE MEANING OF EVOLUTION

A Study of the History of Life

and of

Its Significance for Man

THE MEANING OF EVOLUTION

A Study of the History of Life and of Its Significance for Man

by

GEORGE GAYLORD SIMPSON

YALE UNIVERSITY PRESS, NEW HAVEN AND LONDON

ISBN: 0-300-00952-6 (CLOTH), 0-300-00229-7 (PAPER)

DESIGNED BY BEVERLY KRUGER,
SET IN BASKERVILLE TYPE,
AND PRINTED IN THE UNITED STATES OF AMERICA BY
BOOKCRAFTERS, INC.,
FREDERICKSBURG, VIRGINIA.

20 19 18 17 16 15 14 13 12 11

PREFATORY NOTE TO FIRST EDITION

This book represents, in considerably expanded form, the twenty-fifth series of Terry Lectures, delivered at Yale University in November, 1948; or, strictly, those lectures were a summary of this book. The circumstances and some particulars regarding the Terry Foundation are more especially noted hereafter. I am much indebted to the officers of the University and of the Terry Foundation for this opportunity, and to the officers and staff of Yale University Press for their helpful care and skill in the production of this volume. In addition to special assistance on particular points, acknowledged at the appropriate places in the text, I have profited greatly from the general criticisms of Drs. E. H. Colbert, N. D. Newell, A. Roe, and B. Schaeffer, all of whom have read the whole manuscript (but none of whom is to be understood as endorsing it integrally). With the exception of figure 29, all the illustrations have been specially drawn for this work, mainly by J. C. Germann, with the assistance of N. Altshuler who also drafted some of the nonpictorial illustrations. The index has been prepared by Miss Clara LeVene.

New York City
February 8, 1949

PREFACE TO SECOND EDITION

The original version of this book was written in 1948. Since then, to the astonishment of the author at least, it has gone through many printings and some fifteen editions, ten of them translations from the English into that many foreign languages. Some of these versions were abridged by my hand, but in a way that I now regret because it favored conclusions over evidence. None of the versions has previously been revised aside from the deletions of the abridgment.

All the topics of this work have been subjects of much thoughtful research and searching thought since 1948. Obviously there must now be some points to change and many to add or, at least, to bring up to the moment. It is not obvious but it is charitable to assume also that the author has increased both his knowledge and his wisdom. Under those circumstances there was temptation to write a new book rather than to revise the old. However, this book continues in demand and the choice is not whether to drop it or to write a different book but whether to let it continue becoming gradually obsolescent or to revise it adequately but minimally. The time has come when the latter course is obligatory. In the meantime I have written a couple of the other books that logically followed this one.

In revising I have therefore abstained from making this a different book. I have checked every word and have changed, added, or deleted where really called for, but I have retained all the topics, the sequence, the structure, the point of view, and the great majority of the words. Indeed it was gratifying to find that corrections and additions, involving many lesser points, do not seriously affect the broader and more important aspects of the book. For example in 1948 the oldest well-established ages for rocks were about 2,000,000,000 years, and the oldest fairly well-identified and dated fossils seemed to be about 1,000,000,000 years old. Now both rocks and fossils have been dated at about 3,000,000,000 years. However, it was known in 1948 that dates then roughly established for both rocks and fossils were minimal and that older ones must have existed. The discovery of such older ones confirms and strengthens, not contradicts, earlier conclusions. (Those new dates are still minimal.) It would really have been upsetting and have required radical rethinking if it had turned out that the oldest fossils known were only, say, 1,000,000 years old. No such really radical upset has occurred in this or in the other subjects vital to the theses of this work. We can update, clarify, add significant detail, resolve some doubts, and generally (it is believed and hoped) improve; we do not need to apologize or to abandon an essential conclusion.

Many changes nevertheless have been made. They are of the following kinds:

1. References to later research and discussions have been supplied throughout.

2. When such later work involves definite corrections or necessary additions, the text has been modified or amplified accordingly.

3. Numerous small changes in wording have been made to clarify individual sentences or to bring them fully into the context of 1966.

4. A few discussions, mostly in footnotes, have been shortened or expunged because they have lost urgency or interest since 1948. (However, additions slightly outweigh deletions; this revision is not an abridgment.)

Acknowledgments in the earlier prefatory note are still valid and are again endorsed to the large extent that this edition derives from the first. I am again indebted to Yale University Press, among other things for their willingness to accept a true internal revision, involving resetting the whole book. They have further given more than usual assistance in the whole of the task accomplished. My wife, Anne Roe, continues to merit the original and now renewed dedication of the book to her, not only because she has again helped me with this particular text but also because she has helped me in everything. This revision was written while I was a staff member of the Museum of Comparative Zoology, Ernst Mayr director, and I have profited by the facilities and traditions of that great institution as well as by association with Mayr, A. S. Romer, Bryan Patterson, and many other colleagues at Harvard, although they have not been directly involved in this revision.

The number and subjects of the figures are the same as in the first (full) edition, but several have been revised by me and redrafted by Thomas D. Morin. A new index has been made by Ronald Riddle.

The revising has been done at Los Pinavetes, without benefit of Hollywood.

Los Pinavetes, La Jara, New Mexico
12 July 1966

CONTENTS

ILLUSTRATIONS

PROLOGUE

Most of what follows was written at Los Pinavetes, on the slope of San Pedro Mountain in northern New Mexico. The closing passages were written in Hollywood, California.

Outside my study window at Los Pinavetes rises a forest of stately and ancient ponderosa pines. There is a stream bordered by alders, willows, and mountain cottonwoods. Near by is an aspen grove, where trembling leaves have turned from green to flashing yellow and finally have fallen, as this writing has progressed and as another yearly cycle of life has passed from summer toward winter. In the interspersed glades and fields the wild rose thickets have developed their gorgeous hips and their leaves have turned crimson. A hundred other sorts of flowers have bloomed and gone, scattering seeds which lie waiting to repeat the miracle in another year, as they have for so many ages past.

Animal life, no less than vegetation, is abundant and evident here. By night a bear wanders by and mule deer track the mountain. These are timid folk who know, somehow, that they are marked as prey by man. Not so the many members of the squirrel family, unafraid and given to fits of chattering anger at the intruder: chipmunks, tufted-eared pine squirrels, side-striped ground squirrels, drab but bright-eyed prairie dogs, and the rest. Pack rats, clean and attractive, unlike their distant kin who have given rats a bad name, are busily preparing a winter home in the wood pile. Birds have become less abundant as the weather grew colder, but even after heavy frosts jays and woodpeckers are much in evidence and bands of juncos move through from breeding grounds above to winter range below. Trout swim in the stream and water striders skitter over its surface in the pools. There are a thousand or many thousands of other forms of life, from the eagle far up in the sky to invisibly microscopic organisms deep in the soil.

The landscape is literally packed with life. It is as full of living substance as it well could be.

In Hollywood, too, evidence of life is everywhere. Two-legged mammals jam the streets and all around them rise their strange constructions. The very hills are torn down for their passage and even the plants and animals are theirs.

Los Pinavetes and Hollywood are both parts of nature and both bear testimony to the wonder of life. In the mountain forest is a sample of the beautiful and intricate relationships wrought by blind forces from the matrix of life. Within this matrix has arisen one animal, man, who works no longer blindly, often no longer beautifully, but still with fearful intricacy to make a Hollywood.

How has this all come about? Everyone knows that the city has a history. The general outlines, at least, are familiar: the Indians, the early white explorers, the missions, the Spanish town, the growth of the gaudy American metropolis. Changes are still visibly going on and the city is not the same for years or even for months in succession. The forest and its denizens have a history, too, no less than the city. It covers many millions of years as against the few centuries of the city's history. In fact the two are parts of the same larger history, the history of life. The brief history of the city arises out of the long history of the beasts of the forests and the plains and is the surprising culmination of this.

There are many ways of studying the history of life. The geneticist, raising tiny fruit flies in bottles, and the well driller, piercing thousands of feet of the earth's crust in search of oil, are both contributing to this study. One way of pursuing the study, the most direct way and the one around which all other contributions need to be organized to articulate the whole history, is that of the paleontologist. In his most immediate and narrow occupation, the paleontologist is a student of fossils, the preserved remains of ancient life. He seeks these in the rocks throughout all the lands of the earth, takes them to his laboratory, cleans them for study, compares them with each other, identifies and names their kinds, determines their ages, and finally sets up their associations and historical sequences. As with any other sort of worker, the paleontologist's day-by-day attention is mostly focused on the concrete and more or less routine details of the job in hand rather than on the broad whole of which this job is a small part. When he does turn to these broader aspects, the paleontologist becomes not merely a student of fossils but a historian of life. The historian of life takes not only knowledge of fossils but also a tremendous array of pertinent facts from other fields of earth sciences and of life sciences and weaves them all into an integral interpretation of what the world of life is like and how it came to be so. Finally, he is bound to reflect still more deeply and to face the riddles of the meaning and nature of life and of man as well as problems of human values and conduct. The history of life certainly bears directly on all these riddles and problems, and realization of its own value demands investigation of this bearing.

Turning from professional details and daily bread-winning to consideration of these more fundamental problems requires not only the means of acquiring knowledge but also an opportunity and a stimulus for examining its broader significance. The means on which the present enquiry is based were provided in the first instance by the American Museum of Natural History, which, as fully as any other institution, has brought together the necessary materials for study of all aspects of the history of life. The immediate opportunity and stimulus were given by the Dwight Harrington Terry

Foundation of Yale University. The Foundation provides for the delivery of lectures and the publication of books in which students in the many different branches of learning attempt to set forth such truths in their fields as may particularly bear on human welfare, not immediately material welfare but the basis of this and the higher welfare of the mind and the spirit. In setting up this high goal the cardinal principles fixed by the founder were "loyalty to the truth, lead where it will, and devotion to human welfare." Among the subjects that he specified as most likely to bear on this aim were "all the great laws of nature, especially of evolution."

In accordance with this specification the first Terry Lectures and the first volume published (in 1925) for the Terry Foundation by Yale University Press were by J. Arthur Thomson and were titled *Concerning Evolution.*[1] The years since Thomson wrote on essentially the same subject as that of the present volume have seen such rapid progress of knowledge and such a transformation of ideas that a radically different approach is now possible and required. Thomson felt constrained to devote a considerable part of his work to presentation of proofs of the truth of evolution. This would be a waste of time now. Ample proof has been repeatedly presented and is available to anyone who really wants to know the truth.[2] It is a human peculiarity, occasionally en-

1. One other volume, before the present one, has had evolution as a main theme: *The Universe and Life,* by H. S. Jennings, published in 1933. The title of still another volume, *Evolution in Science and Religion,* by R. A. Millikan, may be misleading in that it refers not to the evolution of life but to the changing ideas of mankind regarding science and religion. One of the most recent volumes, Margaret Mead's *Continuities in Cultural Evolution* (New Haven, Yale University Press, 1964), carries the evolutionary concept into a different sphere but one that is still biological in an extended sense.

2. Among the very numerous sources, reference should still be made to the work that first substantially established this truth: Darwin's *The Origin of Species by Means of Natural Selection* (Murray, London, 1859

dearing but more often maddening, that no amount of proof suffices to convince those who simply do not want to know or to accept the truth. Reiteration for the sake of these wishful thinkers would be futile, and reiteration for those who do want to know the truth is quite unnecessary because they already know it or can easily find it in other works. In the present study the factual truth of organic evolution is taken as established and the enquiry goes on from there.

Thomson also devoted much attention to the proposition that evolution and religion are compatible. Of course there are some beliefs still current, labeled as religious and involved in religious emotions, that are flatly incompatible with evolution and that are therefore intellectually untenable in spite of their emotional appeal. Nevertheless, I take it as now self-evident, requiring no further special discussion, that evolution and *true* religion are compatible. It is also sufficiently clear that science, alone, does not reach all truths, plumb all mysteries, or exhaust all values and that the place and need for true religion are still very much with us. In the following pages much is said about ethics and other subjects with some bearing on religion. This is, however, all said from the point of view of the scientific study of the history of life, which here impinges on religion but remains distinct from it. Of religion, as such, nothing is said in this book, not because of failure to recognize its value and importance, but because that is not my subject.

This book is nontechnical and it is addressed not specifically to other students of the history of life but to anyone

---

[and many later editions]). The first edition, in some respects the best, has recently been reprinted in facsimile by Harvard University Press (Cambridge, 1965) with a new introduction by Ernst Mayr. The interesting changes made by Darwin in subsequent revisions can be followed in the variorum edition by Morse Peckham (Philadelphia, University of Pennsylvania Press, 1959). Among the recent texts that summarize the decisive evidence for evolution is P. A. Moody's *Introduction to Evolution* (New York, Harper and Brothers, 1962).

who has an adult and intelligent interest in this subject. It is hard to see how any intelligent adult could fail to take such an interest. I do not delude myself that the book is easy to read throughout or can be lightly skimmed with much comprehension. The subject is not easy; indeed, there are few that are harder. It would be impossible to make it as easy to read as a murder mystery and still be fair to the reader who really wants to grasp the meaning of the subject for him and who would resent it if it were self-consciously written down to an adolescent level. It would take many times this space and much greater talents to make so hard a subject really easy. This presentation is, however, written in the common language, not in the abbreviated technical jargon of the professionals, and it is not addressed primarily to the professionals.

Some of the professional students of one topic or another here touched on are, of course, likely to read this book in spite of the fact that it is not specifically intended for them alone, and a few brief side remarks to them may be called for. Some of them may be surprised that the truth of evolution and the compatibility of science and religion are taken as requiring no discussion in this enquiry, but that the same attitude is not taken toward vitalism and finalism, issues that seem to some scientists as dead as special-creationism. The fact is that these scientists are wrong and that these issues are not dead. Many who have accepted the truth of evolution, and even some who are well instructed in the facts of the history of life, still insist on vitalistic or finalistic interpretation of its processes. Consideration of these glosses and exposure of their invalidity is still an essential part of enquiry into the meaning of life. Choice between naturalism or materialism and finalism or vitalism lies at the very heart of the problems to which this study is devoted. Omission of this subject would produce a *Hamlet* without a Prince of Denmark in the cast.

On the other hand, it may seem a serious omission, or perhaps a touch of naïveté, that a basic philosophical position is taken without explicit notice. It is assumed that a material universe exists and that it corresponds with our perceptions of it. The existence of absolute, objective truth is taken for granted as well as the approximation to this truth of the results of repeated observations and experiments. That such assumptions are debatable is evident from the violence with which they have been debated at various times. In practice, however, we all have to take it either that they are true or that we necessarily proceed *as if* they were true. Otherwise there is no meaning in science or in any knowledge, or in life itself, and no reason to enquire for such meaning.

As a final aside to the professionals, some of them may be concerned over some of the usages of words, especially of words that have both colloquial and technical meanings. The word "type," for instance, is repeatedly used and some of my colleagues may take it that I am insidiously afflicted with Aristotelian or Platonic idealism or with archetypal metaphysics. Others may suspect some confusion with the technical designation of types in formal taxonomy. No such implications are really present. The word is used in a wholly colloquial sense, in the hope that it will convey a generally correct idea without need for tedious explanation of all the possible technical usages and of the history of related concepts in science and in philosophy.

Specialist and nonspecialist alike may note the scanty attention given to the evolution of plants, which is confessedly a serious omission in a study of the history of life. There is, however, need for reasonable brevity, and despite abundant differences in detail the principles of plant evolution are generally the same as those of animal evolution.[3]

3. An excellent work devoted entirely to plants is *Variation and Evolution in Plants* (New York, Columbia University Press, 1950) by G. Ledyard Stebbins. Verne Grant's *The Origin of Adaptations* (New York,

Man is an animal, so that animal evolution is usually more interesting to him and is also more likely to have meaning for him and to elucidate his place in the cosmos. For the same reasons, discussion of the meaning of evolution for man may properly emphasize the vertebrates among animals and the mammals among vertebrates. It is also fair to add that I have devoted most attention to subjects in which I am most nearly competent or at least with which I have best first-hand acquaintance.

The general plan of the enquiry has three phases which correspond approximately with attempts to answer three questions:

What has happened in the course of the evolution of life?

How has this been brought about?

What meaning has this in terms of the nature of man, his values and ethical standards, and his possible destiny?

The three main sections of this book, between this prologue and the epilogue, are attempts to answer each of these three questions in sequence. The questions are progressive, not independent. Answer to the second requires prior answer to the first. The third, the master question, depends on answers to both the first two. Tremendous questions, these. To think that complete or final answers have been provided would require a measure of conceit impossible to anyone who has spent much time watching the play of life at Los Pinavetes, or even in Hollywood where conceit blossoms more easily. It is only possible to hope that some steps toward comprehension have been made and that these steps have been taken in the spirit prescribed by the Terry Foundation, of loyalty to the truth and devotion to human welfare.

Columbia University Press, 1963) includes animal evolution but is by a botanist and is particularly strong in his special field.

PART ONE

The Course of Evolution

"And our view of the human scene becomes narrow, unillumined, and passionate if we do not rise above its immediate urgency and see it in its cosmic roots and backgrounds."

M. R. Cohen, in *The Faith of a Liberal.*

I. GEOLOGIC TIME

The meaning of human life and the destiny of man cannot be separable from the meaning and destiny of life in general. "What is man?" is a special case of "What is life?" Probably the human species is not intelligent enough to answer either question fully, but even such glimmerings as are within our powers must be precious to us. The extent to which we can hope to understand ourselves and to plan our future depends in some measure on our ability to read the riddles of the past. The present, for all its awesome importance to us who chance to dwell in it, is only a random point in the long flow of time. Terrestrial life is one and continuous in space and time. Any true comprehension of it requires the attempt to view it whole and not in the artificial limits of any one place or epoch. The processes of life can be adequately displayed only in the course of life throughout the long ages of its existence.

The spatial scope of life as we know it is clearly defined as the planet Earth. There may be living things elsewhere in the universe, but if so, they are unknown to us and are of a different stock from ours, or from any familiar or important to us.[1] It is overwhelmingly probable that our particular brand of life arose on our familiar planet and has always been confined to it. This limit in space also sets limits in time. It is clear that this planet had a beginning and that its life arose after—probably long after—that beginning. The time pertinent to our study of life is Earth time, in other words, geologic time.

Fortunately for this enquiry, the crust of the earth contains clocks. We are, to be sure, still in the condition of infants just learning to read time and lucky to hit within an hour or two. Our laborious reading of the geologic clocks still may miss by tens of millions of years. There is, however, assurance that the approximate orders of magnitude are known. The reading of the geologic clocks clearly excludes both the mere thousands of years of earlier Christian cosmogony and the countless billions of still earlier oriental cosmogonies.

The most nearly accurate of the several known sorts of geologic clocks operate by radioactivity. Man has learned, to his own horror, how to produce atomic disintegration at catastrophic rates, but the radioactive elements distributed in minerals in the crust of the earth are far below the critical concentrations necessary for such reactions. As they occur in nature, these elements disintegrate at a deliberate and invari-

1. Since the original version of this book was written (in 1948) there has been tremendous advance in space technology and this has led to increased interest in the possible existence of life elsewhere, a topic now sometimes called "exobiology," although that peculiar science has no known subject matter. It is still true that we know of no life elsewhere than on earth, and (in my opinion) it is highly improbable but not quite impossible that we will have any such knowledge in the foreseeable future. Interesting as the subject is, it would be beside the point to insert discussion of it in the present book. I have briefly considered it in Chapter 13 of *This View of Life* (New York, Harcourt, Brace and World, 1964).

able pace. Uranium, for instance, breaks down into other elements at a rate that will leave about three-quarters of the original uranium unchanged after 2,000,000,000 years. Inert end products of this radioactive decay are the gas helium and lead of a particular sort (an isotope with atomic weight 206). The proportions of uranium and of this sort of lead in a given mineral change slowly but steadily, so that these proportions have a precise and known relationship to the lapse of time since the mineral was formed.

In principle, then, the age of a rock can be determined by this recipe:

Find a mineral that contains uranium and that was crystallized when the rock was formed. Determine the proportion of uranium to that of lead derived from uranium. Apply the formula for rate of radioactive decay of uranium to lead. The answer is the age of the rock in years.

Other radioactive elements and end products can be used in the same way but with different constants in the formula for transmutation. Radioactive potassium and argon derived from it have recently proved to be particularly useful, producing many potassium/argon or K/Ar dates. Minerals datable in this way are more widespread than those containing measurable uranium and its several end products and are more often connected with geological events for which dates are desired. Still another method of dating, also radiometric but depending on quite a different principle, is based on radioactive carbon (C^{14}). Although highly useful to archaeologists (students of human prehistory), its usable range extends back only forty thousand years or so, a negligible period of time to paleontologists and not relevant for our present discussion.

Although these methods are constantly being refined, they still involve many difficulties and inaccuracies. Rocks approximately 3,000,000,000 years old have now been dated, and rocks even older evidently exist. Less direct and less

ROUGH ESTIMATES OF: MILLIONS OF YEARS SINCE BEGINNING		DURATION IN MILLIONS OF YEARS	SEQUENCE TABLE, OLDER BELOW AND YOUNGER ABOVE: ERAS	PERIODS	EPOCHS	SOME FEATURES OF THE LIFE RECORD
			CENOZOIC	QUATERNARY	RECENT	MAN
2		2			PLEISTOCENE	
12		10		TERTIARY	PLIOCENE	MAMMALS AND BIRDS NUMEROUS; FLOWERING PLANTS NUMEROUS; BONY FISHES NUMEROUS
25	65	13			MIOCENE	
35		10			OLIGOCENE	
52		17			EOCENE	
65		13			PALEOCENE	
135		70	MESOZOIC	CRETACEOUS	(EPOCH DIVISIONS NOT NEEDED FOR PRESENT PURPOSES)	
180	165	45		JURASSIC		BIRDS ARISE; MAMMALS ARISE; REPTILES NUMEROUS
230		50		TRIASSIC		AMPHIBIANS NUMEROUS
280		50	PALEOZOIC	PERMIAN		CRISIS IN MARINE LIFE
345		65		CARBONIFEROUS		
405		60		DEVONIAN		AQUATIC VERTEBRATES NUMEROUS, RISE OF TRUE FISHES
425	370	20		SILURIAN		RISE OR SPREAD OF LAND ANIMALS AND PLANTS
500		75		ORDOVICIAN		FIRST KNOWN VERTEBRATES; ALL BASIC TYPES OF AQUATIC ORGANISMS APPEAR
600		100		CAMBRIAN		FIRST ABUNDANT FOSSILS
MORE THAN 3,000	MORE THAN 2,400		PRECAMBRIAN	(PERIOD DIVISIONS NOT WELL ESTABLISHED)		RELATIVELY FEW, PRIMITIVE FOSSILS

IN THE UNITED STATES, BUT NOT ELSEWHERE, THE CARBONIFEROUS IS SOMETIMES DIVIDED INTO AN EARLIER MISSISSIPPIAN PERIOD AND LATER PENNSYLVANIAN PERIOD.

Fig. 1. The time scale of earth history and of life.

reliable evidence suggests that the whole age of the earth as a separate entity (planet) may be very roughly on the order of 4,500,000,000 years, more economically written as 4.5×10^9 years. These figures are staggeringly large, and yet they are finite and set an end to wandering in the apparently infinite range of cosmic time. The earth more or less as we know it, capable of occupation by some form of life, has endured for more than 3×10^9 years.

The succession of events in earth history has been established more accurately than their dates and durations in years. It is, as yet, more useful and customary for students of the history of life to say that a given animal lived, for instance, in the Permian period than to say that it lived perhaps 250 million (250×10^6) years ago. We know absolutely that a Permian animal lived after its Carboniferous forebears and before its Triassic successors, even though we may have serious doubts just how many years, before and after, were involved. In discussion of the course of evolution some of these geologic period and epoch names must be used. Those necessary for this account are given in the accompanying table (fig. 1) and they are listed and discussed in more detail in any book on historical geology.[2] Besides the 3×10^9-year round figure for the oldest rocks yet dated, a number of fairly accurate year dates have been obtained corresponding with other parts of the geologic table of periods and epochs. It is thus possible to arrive at estimates for the year ages and durations of all the various periods and epochs. Estimates by good authorities still may differ by as much as two to one for year durations of some of the periods, but discrepancies in well-grounded recent estimates are usually less than that. Rough as these approximations are in the present state of knowledge, they give some idea of the durations of various

2. For example, B. Kummel's *History of the Earth* (San Francisco, Freeman, 1961).

events and the rates of some of the processes of the evolution of life.[3]

3. Many of the available radiometric dates are listed and discussed in J. L. Kulp, "Geologic Time Scale" (*Science, 133* [1961], 1105–1114) and J. F. Evernden, G. H. Curtis, D. E. Savage, and G. T. James, "Potassium-Argon Dates and the Cenozoic Mammalian Chronology of North America" (*Amer. Jour. Science, 262* [1964], 145–198). I have used those data, with some modifications, in compiling fig. 1.

II. THE BEGINNINGS OF LIFE

The origin of life was necessarily the beginning of organic evolution and it is among the greatest of all evolutionary problems. Yet its discussion here will be brief, almost parenthetical. Our concern here is with the record of evolution, and there is no known record bearing immediately on the origin of life. The first living things were almost certainly not apt for any of the usual processes of fossilization. It is unlikely that any preserved trace of them will be found or recognized. Indeed it is improbable that the discovery of such remains, if any do exist, would greatly advance knowledge of how life originated. At this lowest level little could be learned from the preserved form: the problem is physiological, not morphological, and it seems that form must develop above the molecular level before it can serve as a particularly useful clue to function.

Improbable as it is that paleontology will contribute significantly, the problem is open to study. Recent work in biochemistry and especially studies of cell structure and physiology are converging hopefully on this mystery.[1] The solution has not yet been reached and it may be near, or distant. Yet these studies show that there is no theoretical

1. Publications on the origin of life are extremely numerous. The following are a few of the more general works: A. I. Oparin, *The Origin of Life* (London, Oliver and Boyd, 1950) and (as editor) *The Origin of Life on Earth* (London, Pergamon Press, 1960). G. Ehrensvärd, *Life: Origin and Development* (Chicago, University of Chicago Press, 1962). M. Florkin, editor, *Some Aspects of the Origin of Life* (London, Pergamon Press, 1961). S. W. Fox, editor, *The Origins of Prebiological Systems*

difficulty, under conditions that probably existed early in the history of the earth, in the natural origin of complex molecules capable of influencing or directing the synthesis of units like themselves. It has been demonstrated experimentally that nominally organic molecules, those containing carbon in combinations of varying complexity, can originate by nonorganic (nonliving) means. Thus basic building blocks of life, notably nucleic and amino acids, arose, and hence eventually DNA (deoxyribonucleic acids), now known to code and induce the formation of specific proteins, and the catalytic and other proteins themselves: long, linked sequences of the primitive amino acids. Once such conjoined or interacting molecular systems had acquired the properties of inducing replication and of variation of detailed sequences in the replicas, natural selection would inevitably begin to operate. Natural selection in this connection means simply that some variant systems would produce more replicas—would reproduce more efficiently—than others, and that this differential continuing through thousands and millions of generations would lead steadily to increasingly complex, even more effective reproductive populations.

There is no reason to postulate a miracle. Nor is it necessary to suppose that the origin of the processes of reproduction and mutation was anything but naturalistic. That is, study of these basic functions in existing organisms indicates that their novelty lay only in the organization or state of matter and its surroundings, and not in the rise of any new property or principle, physical or nonphysical. Once this point is established the origin of life is stripped of all real

(New York, Academic Press, 1965). George Wald, "The Origins of Life," (*Proc. Nat. Acad. Sci.*, *52* [1964], 595–611).

There are of course also a great many highly technical studies of biochemical detail, among which I cite just one as example: C. Ponnamperuma and R. Mack, "Nucleotide Synthesis under Possible Primitive Earth Conditions" (*Science*, *148* [1965], 1221–1223).

mystery, regardless of whether it proves possible in a brief time in a modern laboratory to repeat the event that occurred in the course of millions of years when the earth was young.

Above the molecular level, the simplest fully living unit is almost incredibly complex. It has become commonplace to speak of evolution from ameba to man, as if the ameba were the simple beginning of the process. On the contrary, if, as must almost necessarily be true, life arose as a simple molecular system, the progression from this stage to that of the ameba is at least as great as from ameba to man. All the essential problems of living organism are already solved in the one-celled (or, as some now prefer to say, noncellular) protozoan, and these are only elaborated in man or the other multicellular animals. The step from nonlife to life may not have been so complex, after all, and that from cell to multicellular organism is readily comprehensible. The change from replicating molecule to protozoan was probably the most complex that has occurred in evolution, and it may well have taken as long as the change from protozoan to man. In the more than 3,000,000,000 years of the history of life, time is available even for this.[2]

All this, however, is somewhat by the way, only a prelude to the recorded history of evolution. Still the prelude is not quite over, for the curtain did not rise all at once on this drama. Although the rocks of 3,000,000,000 years and more are available for our inspection, only those formed during the last fifth or so of this time contain reasonably satisfactory evidence of life. Some early Cambrian rocks, laid down about

2. The enormous complication of even what we consider "simple" cells, the now irreducible unit of truly living systems, is exemplified in *The Scientific Endeavor* (New York, Rockefeller Institute Press, [not dated but issued in 1964]), especially the chapters by G. E. Palade, E. L. Tatum, and T. M. Sonneborn, and in J. A. Moore (editor), *Ideas in Modern Biology* (Garden City, Natural History Press, 1965), especially the chapters by E. De Robertis and K. R. Porter.

600,000,000 years ago, are crowded with fossils. One place or another on earth there are also rich fossil deposits of almost all ages since the early Cambrian. But in rocks earlier than the Cambrian, representing the great span of upwards of 2,500,000,000 years, fossils are generally lacking and even when present may be difficult to interpret.

The search for Precambrian fossils has been intense and often disheartening. Few traces have been found, and of those few some later proved not to be organic or not to be Precambrian. *Eozoon,* proudly named "the dawn animal," is now considered to be no animal at all, nor yet a plant or any form of life but a mere inorganic precipitate. Other claimed discoveries, especially of some rather highly organized jointed animals, are too doubtful as to age to be accepted as definitely Precambrian. One ingenious claim relates to structures named *Corycium* found in some numbers in early Precambrian rocks in Finland. Analysis of carbon from these objects reveals that it includes two forms, the isotopes C^{12} and C^{13}, in ratios now characteristic of organic rather than inorganic concentrations.[3] Unfortunately *Corycium* is a nondescript sort of globule and we are still wholly in the dark as to what sort of plant or animal it may be, if any. At any rate this study strengthens the old speculation that occurrences of carbon (usually altered to graphite by heat and pressure) in Precambrian rocks may derive from otherwise unrecognizable organisms.

3. K. Rankama, "New Evidence of the Origin of Pre-Cambrian Carbon," (*Bull. Geol. Soc. Amer., 59* [1948], 389–416). Somewhat analogous graphitic bodies have since been found in Canada: B. L. Stinchcomb, Harold L. Levin, and Dorothy J. Echols, "Precambrian Graphitic Compressions of Possible Biologic Origin from Canada," (*Science, 147* [1965], 75–76). Organic compounds, without preserved organic shape (morphology), have also been reported from a number of Precambrian rocks, for example, J. Oró, D. W. Nooner, A. Zlatkis, S. A. Wikstrom, and E. S. Barghoorn, "Hydrocarbons of Biological Origin in Sediments about Two Billion Years Old," (*Science, 147* [1965], 77–79).

There are, however, definite fossils in Precambrian rocks, and some of these carry the recorded history of life back to the almost incredible antiquity of approximately *three billion* years b.p. (before present—the usual formula for geological dates in years). Somewhat similar occurrences are also known from later dates, roughly two and one billion years b.p.[4] These fossils consist of single cells and simple aggregations of cells. As would be expected of such extremely old organisms, they cannot be exactly placed in any groups still alive today, but they resemble bacteria and blue-green algae, confirming the view that these are the most primitive fully developed living organisms.[5]

From sometime around one billion b.p. onward there was also considerable development of algal reefs, limestone deposits formed by multicellular masses of seaweeds.[6] In spite of numerous earlier claims, unquestionable multicellular animals are not now known earlier than the latest Precambrian, indeed in beds that are not certainly older than earliest Cambrian.[7]

4. E. S. Barghoorn and J. W. Schoff, "Microorganisms Three Billion Years Old from the Precambrian of South Africa," (*Science, 152* [1966], 758–763). E. S. Barghoorn and S. A. Tyler, "Microorganisms from the Gunflint Chert," (*Science, 147* [1965], 563–577). J. W. Schopf, E. S. Barghoorn, M. D. Maser, and R. O. Gordon, "Electron Microscopy of Fossil Bacteria Two Billion Years Old" (*Science, 149* [1965], 1365–1367). E. S. Barghoorn and J. W. Schopf, "Microorganisms from the Late Precambrian of Central Australia," (*Science, 150* [1965], 337–339).

5. That role is sometimes ascribed to viruses, which are very much simpler systems including nucleic acids. They probably do give some hints about precellular, molecular stages in the origin of life, but they are not now believed to be actual survivors from that stage or to be truly alive in the full sense of the word.

6. C. L. Fenton and M. A. Fenton, "Pre-Cambrian and Paleozoic Algae" (*Bull. Geol. Soc. Amer., 50* [1939], 89–126).

7. M. F. Glaessner and B. Daily, "The Geology and Late Precambrian Fauna of the Ediacara Fossil Reserve," (*Records S. Australian Mus., 13*

Although we are thus finally acquiring a spotty record of life far back into the Precambrian, it remains true that Precambrian fossils are excessively rarer than those from the Cambrian onward, even in rocks that seem equally suited to preserve traces of life. It is unlikely that the scarcity of Precambrian fossils is entirely due to their greater antiquity.

This major mystery of the history of life has naturally excited a great deal of argument and speculation. Some students have supposed that the groups of animals whose remains appear in the Cambrian suddenly came into existence at about that time. The generally held, more orthodox theory is that these groups really arose slowly and well back in the Precambrian but that those earlier forerunners had no shells or other hard parts. As a rule, only hard parts are preserved as fossils. Conditions permitting preservation of soft parts do occasionally exist—there are some remarkable examples from the Cambrian, itself, as it happens—but they are so exceptional that the absence of such deposits in the Precambrian would not be surprising. As to how it happened (if it did) that Precambrian animals did not and their Cambrian descendants did have hard parts, this is subject for further speculation. At least, as speculation it is not unreasonable that this might have occurred.

It is possible to combine the two main theories in somewhat modified form, and the combination seems more likely than either theory alone. That seven or eight major grades of animal organization originated instantaneously and simultaneously is quite incredible, but it is both possible and probable that their various origins may have occurred within a span of some millions of years and may have been accelerated, occurring at evolutionary rates far greater than any in their later histories. It is also nearly incredible that some groups of ani-

---

[1959], 363–401). M. E. Taylor, "Precambrian Mollusc-like Fossils from Inyo County, California," (*Science, 153* [1966], 198–201).

mals (such as crustaceans), which depend on hard parts for their functioning, arose as such before they acquired hard parts. It is, again, quite incredible that many entirely distinct groups of animals stumbled, so to speak, on the invention of hard parts at exactly the same time and by pure coincidence. But it is probable that development of hard parts and of other diagnostic characters of groups distinguished by these occurred as parts of a single, rather rapid evolutionary process; and that the development of hard parts by different groups at very approximately (by millions of years) the same time was not pure coincidence.

The simultaneity of these events has been exaggerated by the too sweeping statement that most of the animal phyla known as fossils appear in the Cambrian. The Cambrian was the longest single period of those into which geologists divide the last 600,000,000 years of earth history. Its length is now estimated at from 90,000,000 to 100,000,000 years. Even the early part of the period had a duration probably not less than 30,000,000 years. These are long times, even to a geologist, and a great deal of evolution could occur in them, even at moderate rates of evolution. Thirty million years ago your ancestor was something like a primitive ape and 60,000,000 years ago something like a tree shrew. The various Cambrian animal phyla do not all appear as fossils in the very earliest rocks of that period but they straggle in throughout its earlier part, or later. As a whole, the early Cambrian representatives of the groups that did appear then are markedly simple and generalized, as if near the origin of their respective lines.

Relatively simple, small, soft-bodied animals of several basic types and myriad detailed sorts had probably existed in the Precambrian. Early in that time the marvels of the origin of life and of the development of the cell and cellular organism had been wrought. Toward the end of Precambrian time stages had been reached which had high plasticity and great potentialities. A crisis of rapid evolution, of divergence, of specialization for new or different major ways of life was at hand. This may have been triggered by crucial changes in the atmosphere, which may at about this time have reached a concentration of oxygen favoring animal respiration but still admitting enough dangerous radiation to make shells and dermal shielding advantageous. Once the possibility existed, the event would tend to be rapid. It is one of the best attested generalities of evolution that its rate is exceptionally fast when an evolving group takes on some hitherto unexploited way of life or adaptive zone. Interaction among the various groups involved would tend to intensify their divergence and accelerate its speed. Development of hard parts would permit, demand, and hasten structural change and would be selectively advantageous both as defense or offense and as a means of developing new activities and invading new environments. Given the chemical and physical possibilities in the environment and the physiological (genetic) possibility in the organisms, such a development in some groups would cause selective pressure for its development in others also, whether as defense against newly armed aggressors, as aid in aggression against newly protected prey, or as means of competition for common needs with those similarly equipped.[8]

So, speculatively but not groundlessly, may be explained the first and perhaps the greatest crisis in the recorded history of life, the crisis that begins the continuous record. Regarding

8. An excellent review of this problem before the most recent discoveries of very ancient Precambrian fossils was provided by P. E. Cloud, Jr., in "Some Problems and Patterns of Evolution Exemplified by Fossil Invertebrates," (*Evolution, 2* [1949], 322–350). I also discussed it at some length in my chapter (pp. 117–180) in S. Tax (editor), *Evolution after Darwin*, vol. 1 (Chicago, Univ. of Chicago Press, 1960). Possible connection with atmospheric change, with references to earlier work, is considered by A. G. Fischer in "Fossils, Early Life, and Atmospheric History" (*Proc. Nat. Acad. Sci., 53* [1965], 1205–1215).

the long prehistory of the Precambrian, the principles of evolution must have been the same then, different as were the organisms through which they were working. The paucity of earlier records is bitterly regretted, but there is no reason to fear seriously that what may be learned from the later record about the grand processes of evolution is untrue as regards the less well-known earlier parts of the history of life.

III. MAJOR OUTLINES OF THE FOSSIL RECORD

There are more than 1,000,000 different kinds, species, of animals in the world today.[1] About three-fourths of these are insects. Insects did not appear until the early Carboniferous and they are probably more varied and abundant today than at most, perhaps any, times in the past. Land life in general began modestly in the Devonian (or perhaps in cryptic form somewhat earlier) and must, even apart from insects, be more or less near a maximum now. In the earlier records there are remains only of aquatic, mainly marine, life. As far as we know, only the seas have been continuously inhabited since Cambrian (by inference also Precambrian) times. They may at times have exceeded their present approximately 150,000 species, but probably not by much. In general it is clear that the number of species existing at any one time has increased greatly, perhaps fivefold or even tenfold, from the Cambrian to the present time. Estimates of the numbers of individual animals stagger the imagination and are so very

1. Authoritative estimates for all the major groups are given by Lord Rothschild in *A Classification of Living Animals* (London, Longmans, 1961).

In all the following discussion, reference is made to the categories of what is called the Linnean hierarchy of classification. The major subdivisions of the animal kingdom are the phyla (singular, phylum). Successively lesser categories, each included in the preceding higher category, are classes, orders, families, genera (singular, genus), and species (singular also species, not specie). Finer subdivisions are obtained by prefixing "super-" and "sub-" to these terms.

approximate as to be almost meaningless as figures, but it is safe to conclude that these numbers too have increased markedly since the Cambrian.

This increase in the total variety and abundance of animal life in the world has not been continuous or regular. There have been times when increase was particularly rapid, others when it was very slow, and even periods when there seems to have been a reversal and the total of animal life decreased temporarily. There has, nevertheless, been a general and average increase. This broad tendency has not affected the different general types of animals in any uniform way, even as broad averages over their whole spans. Their records fluctuate and differ in remarkable and instructive ways. Some types have risen rapidly and then dwindled slowly to extinction. Others have slowly attained importance, then quickly declined. Still others have continued from ancient times with relatively little fluctuation. Within general types, too, the numerous subtypes have had similarly varied histories. All these and yet other patterns of rise and fall may be richly exemplified from the record.

The bewildering array of tens of millions of minor species of animals, ancient and recent, tends to obscure the broader pattern of life history. Endlessly diverse as these groups seem to be, they represent variations on a relatively small number of basic themes, general types of organization. The broadest of these themes are those formalized by zoologists as phyla, each of which represents a distinct plan or level of anatomical organization, with its attendant possibilities in the functioning of the animals. Put in another way, it may be said that at the base of each phylum lies the acquisition of some distinctive complex of characters which happened to be rich in evolutionary potentialities worked out, often in strikingly diverse ways, in the later members of the group.

Most zoologists place about twenty groups of animals at the major level and call them phyla, although some would raise thirty or more to this grade of classification and others give such high rank to fewer than twenty groups. The differences of opinion relate almost entirely to a number of peculiar, small, soft-bodied living animals of no great importance in the modern fauna and without known fossil records. If we knew their histories, we would probably find that most of these represent some relatively unsuccessful, peculiarly aberrant development from one of the better established major phyla. Interesting as these are as curious developments of the enormous potentialities of life, their part in the great panorama is negligible. The broader outlines of the history of life involve mainly the following phyla:[2]

Protozoa. These animals have no differentiation of their substance into separate cells. They carry on the life processes with relatively simple, although already very complex, orga-

2. More detailed descriptions of the various phyla and pictures of their members are given in many readily available books on zoology. For invertebrates at a nontechnical level no one has excelled R. Buchsbaum's *Animals without Backbones* (Chicago, Univ. of Chicago Press, 1961). A book on vertebrates in similar style is A. S. Romer's *The Vertebrate Story* (Chicago, Univ. of Chicago Press, 1959). Among many other books, see also J. Z. Young, *The Life of Vertebrates* (Oxford, Clarendon Press, 1950), and *The Life of Mammals* (Oxford, Clarendon Press, 1957).

A brief and elementary account of paleontology and the various groups of fossil plants and animals is given in my *Life of the Past* (New Haven, Yale Univ. Press, 1953). These subjects are also reviewed in a broad biological perspective in G. G. Simpson and W. S. Beck, *Life. An Introduction to Biology* (New York, Harcourt, Brace and World, 1965). The standard modern work in English on fossil vertebrates is A. S. Romer, *Vertebrate Paleontology* (Chicago, Univ. of Chicago Press, 1966). Other good books on fossil animals, especially invertebrates, include W. H. Easton, *Invertebrate Paleontology* (New York, Harper and Brothers, 1960), and J. R. Beerbower, *Search for the Past* (Englewood Cliffs, Prentice-Hall, 1960). A monumental, multivolume, highly technical *Treatise on Invertebrate Paleontology,* edited by R. C. Moore, is in course of publication by the Geological Society of America and the University of Kansas Press. On fossil plants see, for example, H. N. Andrews, *Ancient Plants and the World They Lived In* (Ithaca, Comstock, 1947).

nization within an undivided mass of protoplasm analogous to a single cell of the animals of other phyla, all of which are multicelled or metazoan. Such organization can function only in small animals, and most of them are microscopic. In spite of their relative simplicity many of them secrete hard, supporting structures, frequently complex and beautiful, and these types, particularly the Foraminifera, are common as fossils.

Porifera. These are the sponges. Their bodies have many cells and some of the separate cells are differentiated in form and physiology, but they are not clearly arranged in definite layers or well divided into masses or tissues of special function. Many sponges have skeletons of lime, silica, or spongin (the material of a bathroom sponge), and these have left a long fossil record, although this is not particularly continuous or impressive.

Coelenterata. This is a highly varied phylum of aquatic, mainly marine, animals the basic character of which is primary differentiation of the body cells into two definite tissue layers. Most of them have more or less cylindrical bodies with radial symmetry, although this is commonly modified in various ways. They often have limy skeletal supports and the most familiar group, that of the corals, has left one of the best fossil records known to us.

Graptolithina. This extinct group is classed as a phylum not so much because its peculiarities seem really to represent a basic structural theme as because they are hard to interpret, in the absence of living representatives, and could represent a development from any of several other, better established phyla, such as the Coelenterata or the Chordata. They were like many coelenterates and a few primitive chordates in being colonial, with numerous individuals secreting a common skeletal support and living in organic continuity. Their fossil record, all we know of them, is excellent, much as we miss knowledge of their vanished soft tissues.

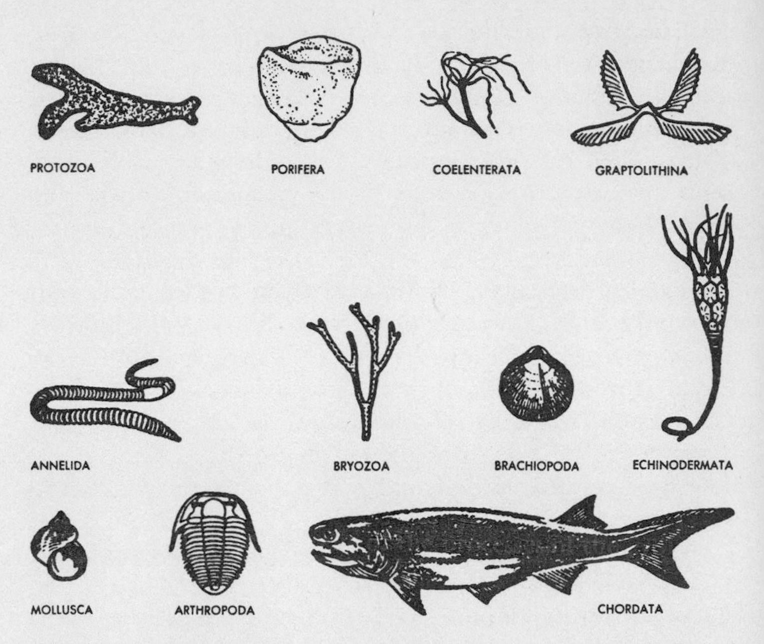

Fig. 2. Examples of the most important basic types, phyla, of animals. All the phyla include large numbers of extremely diverse animals many of which look radically different from these examples. For characterizations of each group see the text. The specimens are drawn to different scales.

Platyhelminthes, Aschelminthes, Annelida. These are mainly "worms" in popular conception, but their basic anatomical characters are so varied that their separation into several phyla is well justified. Indeed students of the living forms commonly recognize other, but less important, phyla of wormlike animals. Collectively, the worms are now varied, abundant, and extremely important parts of the economy of nature, particularly the Platyhelminthes, flatworms, Aschelminthes, mostly nematodes or roundworms, and Annelida, jointed worms (including the familiar angleworms). They have doubtless had this importance from very ancient times, since the Cambrian at least, but they are small, soft-bodied animals and the fossil record is lacking for most of them, poor for the rest. This is probably the greatest deficiency of the record. There are a few fossils, especially of the annelids, most complex of the wormlike groups, but these are insufficient to give much useful information as to their evolution.

Bryozoa. These are colonial, aquatic, mostly marine, animals which broadly resemble the graptolites and some of the colonial corals. They are, however, basically more complex than the corals (we cannot be sure about the graptolites) in having three well-differentiated primary tissue layers and a body cavity (coelom) within the intermediate layer. The colonies secrete skeletal supports, one type of which is the familiar "sea moss." These supports are easily fossilized and the record is unusually good for this group.

Brachiopoda. These forms, the marine lamp shells, are highly varied in detail but relatively uniform in basic plan. They are like the clams in secreting a pair of enclosing shells, usually hinged at one end, and their anatomical complexity is similar. The symmetry of the shell is different, however, as each one of the pair is (usually) symmetrical around a midline, and the internal anatomy has a different plan. The surviving forms are few and relatively unimportant in the life of the present seas, but the phylum was of prime importance in earlier, especially Paleozoic seas. Their fossil record is one of the best, thanks to their abundance and their easily preserved shells.

Echinodermata. This is an unusually large and varied marine phylum, with such diverse representatives as the starfishes, sea urchins, and sea lilies. They have advanced tissue and organ differentiation comparable in grade to that of the shellfish, brachiopods and molluscs, although quite different in kind. They have a basic bilateral symmetry which (in adult forms) is commonly overshadowed by an apparent five-rayed radial symmetry. Stalked, attached types, armed types, and globular types moving on spines have repeatedly developed. The middle tissue layer develops a complex, internal, skeleton consisting of separate limy plates. The fossil record is long and rich.

Mollusca. This is the most varied, dominant, and successful of the primarily aquatic phyla, both at the present time and throughout most of the fossil record. Representatives are very familiar: clams, snails, octopuses, and others. The success and variety of the group is shown by the fact that it not only dominates in the sea but is also abundant in fresh waters and has developed successful nonaquatic types, such as the garden snails. The fossil record of most subdivisions is excellent, probably more nearly complete than for any other phylum.

Arthropoda. If sheer weight of numbers is the deciding factor, this is the most spectacularly successful of all the phyla, for it includes the insects. Its basic plan involves the advanced tissue differentiation and jointed structure also seen in the annelid worms, plus the development of legs and of a hard coat or external skeleton. The potentialities are enormous and arthropods occur exuberantly, from the depths of the sea to the upper reaches of the air. Besides the insects, familiar examples are crabs and other crustaceans and the scorpions, spiders, and their allies. There are many known fossil arthropods and the record is good for several groups, although it is deficient for the insects, in proportion to their probable abundance and importance in later geologic times.

Chordata. This phylum shares complex tissue and organ differentiation, bilateral symmetry, and some other features with several preceding phyla, but adds as its most general distinction an internal supporting structure consisting of a longitudinal rod along the back, jointed in all but a few of the relatively simplest types. The jointed forms, which thus comprise most of the phylum, are the Vertebrata. In evolu-

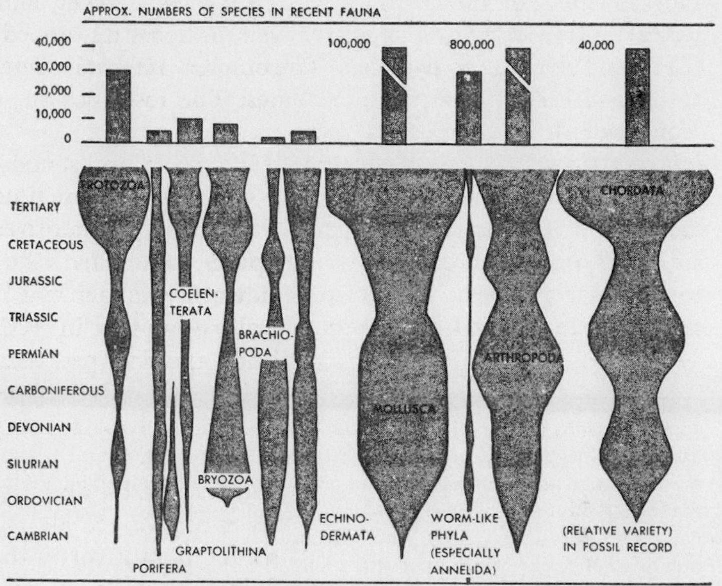

Fig. 3. The broad outlines of the historical record of life. In the lower figure the phyla of animals are represented by vertical bands or pathways the widths of which are approximately proportional to the *known* variety (especially in terms of genera) of the phylum in each of the geological periods since the Precambrian. The upper figure represents the approximate variety (in species) of each at the present time.

tionary success, this plan vies with that of the arthropods—we humans readily assign it first place, for we are chordates ourselves. The record is generally good and is, to us, the most interesting of all. With an anthropocentricity which will be given such defense as it may need later in this book, special attention will be given to the chordates throughout this discussion.

In figure 3 the relative variety of these major phyla today

and the extent of their fossil records are shown.[3] This is a picture, in broad strokes, of what is known of the history of life for the last 500,000,000 years. The notorious imperfection of the fossil record, and the less publicized but existent shortcomings of the students of that record, keep the diagram from representing very closely the picture of life as it really existed. The facts and our knowledge of them unfortunately are not coextensive, here or in any other field of science. Yet these imperfections could easily be exaggerated, and they have been by some overmodest paleontologists. It can fairly be claimed that this and following diagrams of similar nature

3. This diagram and those of figures 5–6, 10–14, and 16 are necessarily approximations only and involve various uncertainties and crudities. Full data of this sort for all groups of fossils have never been compiled and the data that are available are not always properly comparable from group to group, even within a phylum. The diagrams are, nevertheless, based on carefully evaluated data and not on unnecessary approximation or wholly subjective impressions. The widths of the patterns for the different groups, given as proportional to their known variety at the times shown, are based for the most part on data as to the numbers of genera, occasionally of subfamilies or families, recognized by authorities on the group or listed in standard compilations. In dealing with fossils numbers of genera and higher categories give a much better idea of the picture than would numbers of species. Many of the data for invertebrate groups were compiled by N. D. Newell. Among those who assisted Dr. Newell and me in gathering this information are H. Thalman, W. S. Cole, M. L. Thompson, and A. S. Campbell for Protozoa, J. W. Wells for Coelenterata, R. C. Moore for Echinodermata, A. K. Miller and O. Haas for Mollusca, B. F. Howell for Annelida, and A. Petrunkevitch for Arthropoda. Data for the Chordata were mostly compiled from A. S. Romer's *Vertebrate Paleontology* and from my tabulations for mammals. The compilations were made in the late 1940's. Comparable later data are not available. The known or recognized numbers of genera of fossils have increased in all groups in the last twenty years, but the proportional variety, as shown in the figure, seems not to be significantly different. Estimates of numbers of living species have increased, especially for the Protozoa, Mollusca, and wormlike phyla, and the figure has been modified accordingly.

are strongly suggestive, at least, of the general course of events in the history of life. Some of the deficiencies can be discounted with fair assurance. For instance, it is certain that the later part of the pattern for Arthropoda should flair out much more than it does; the present great dominance (in variety) of the insects over any other group of animals must already have existed throughout the Tertiary, and probably still earlier. Similar corrections can often be made in more detail, as will later be exemplified in discussing the Primates.

All these phyla are seen to be of great antiquity. Most of them begin in the Cambrian although, as has already been emphasized, the Cambrian was a very long period so that first appearances of Cambrian phyla are really scattered over some millions of years and not as simultaneous as so generalized a diagram makes them appear. There is little logical order in time of appearance. The Arthropoda appear in the record as early as do undoubted Protozoa, although by general consensus the Protozoa are the most primitive phylum and the Arthropoda the most "advanced"—that is, structurally the most complicated—among the nonchordates (or invertebrates, as all the phyla other than Chordata are often called). Corals and bryozoans do not appear until the Ordovician, although they are less highly organized than many groups that do occur in the early Cambrian.

For the present this failure of the phyla to appear in the order which would be expected as "natural" on the basis of increasing complexity is merely stated as a fact. This peculiarity is not, as a rule, seen among groups within single phyla, which do usually appear in the record in the order of their increasing complexity, specialization, or "progressiveness." The various classes of chordates, for instance, will be shown to have appeared in the record precisely in the sequence that would have been predicted on the basis of their relationships and increasing divergence from or advance beyond a prototype. The phylum Chordata is itself an exception among the phyla, for it does follow what we feel should be, but what is not, the rule. It is to be considered in some sense the most progressive of all the phyla, and it is indeed the last to appear in the record, although not much antedated by the lowly bryozoans.

The diagram bears out the previous statement that the variety of life as a whole has tended to increase since the early Cambrian, although with many irregularities. The same is true of some of the single phyla, but here with such irregularities as almost to outweigh the rule. Most of the phyla do appear with relatively few different types and build up only slowly to any considerable variety. This may seem so natural as to be almost inevitable, and yet it will be shown later that it is not true of the record of many of the lesser groups included within the phyla. That it is true of most, perhaps all, of the phyla well recorded as fossils is another indication that they may really have arisen not long before the time when they enter the record, rather than far back in the Precambrian as often supposed. The Bryozoa constitute the nearest thing to an exception among these well-recorded phyla, for they built up nearly to an all-time maximum in the Ordovician, the period in which they are first known.

Arthropods and chordates have increased, with fluctuation, nearly or quite to the present time and are probably now, or have recently been, more varied than at any earlier time. The same may be but is less clearly true of protozoans, coelenterates, and molluscs, surprising as this may seem for such supposedly lowly groups. On the other hand, the brachiopods are certainly far less numerous now than they had been during most of their long previous history and the bryozoans and echinoderms have probably been more varied and abundant at some time in the past than they are now.

Both these latter groups show two well-separated times of expansion and this points attention to a peculiarity shown, although not always so clearly, in the records of all the existing phyla. In each case the record shows some degree of expansion during the Paleozoic followed by a restriction in the Permian (last period of the Paleozoic) or Triassic (first of the following era) or both. This restriction is invariably followed by another expansion in the Jurassic or Cretaceous.[4]

There is little doubt that this was a real phenomenon for marine life and reflects a Permian-Triassic crisis in the life of the seas. It is doubtful, perhaps nonexistent for land life, although it may have affected land animals, too, in some way. The recorded variety of insects declines in the Triassic, but the record is very inadequate. The greatest decline of the amphibians was toward the end of the Triassic and perhaps too late to be part of the general Permo-Triassic reduction, while the reptiles held their own through this time of crisis only to decline rapidly at a much later date, in the late Cretaceous. Other forms of land animals were either not yet in existence or are too poorly recorded for useful study of this point.

This major crisis suggests a radical and widespread alteration of living conditions on the earth, and there is other independent evidence that such a change did occur during the long span of the Permian and Triassic. The Permian was a period of great mountain building and of climates more sharply zoned and more rigorous than had prevailed since the Precambrian. Glaciation occurred widely in the Permian, and it had not during the long earlier periods of more uniform climates in the Paleozoic. During most of the Triassic the lands still stood exceptionally high above the restricted seas, and climates were still apparently rigorous in many places, although not widely glacial. It is uncertain exactly

4. This Permian or Triassic restriction is not really clear in the record for the Porifera, but the record is too poor to be reliable. The later expansion is not present for the graptolites, but for the excellent reason that their Permian restriction proceeded to zero: they then became extinct.

how these conditions affected marine life, but the association of unusual physical conditions with a crisis in evolution is not likely to be pure coincidence.

This apparent correlation between physical conditions on the earth and the history of its life is a striking example of the unquestionable fact that life and its environment are interdependent and evolve together. Other large-scale examples could be cited and many more have been claimed, but often on doubtful grounds. The fact is that the relationship is so intricate and the effects of changes often so local or so special that really world-wide results from physical changes of the earth seem to have been unusual and are difficult to demonstrate. For instance, the great glacial epoch from which we are now just emerging had surprisingly little effect on the animals that live in warm oceanic waters. More local physical events may, indeed, often be shown to have had wide effects on life, especially as results, direct and indirect, of changes in its distribution. For instance, the rise of the Isthmus of Panama led to very striking changes in the land life of South America, to less extent also in that of North America, by permitting intermigration and mingling of the previously separate faunas of the two continents. Discussion of this and similar events would, however, run ahead of the story here being considered, for their over-all effect on the broadest outlines, on the history of animal phyla, rather than of lesser groups, is almost negligible.

Examination of the make-up of various phyla before and after the Permo-Triassic crisis reveals in several of them striking special cases of a widespread evolutionary phenomenon: replacement, within a given group or adaptive type, of one sort of organism by another. The echinoderms, for instance, illustrate this in an effective way. During the Paleozoic this phylum had prospered and had been represented mainly by sorts known as cystoids, blastoids, and crinoids, of which the latter were particularly successful dur-

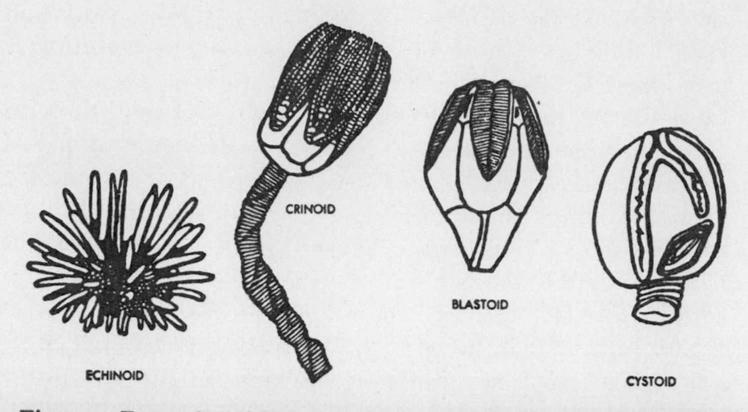

Fig. 4. Examples of four main types of echinoderms. Each group includes widely different sorts of animals, but those shown are fairly typical. The blastoids and cystoids are extinct and the specimens shown lack some features they had when alive. Enlargements are different for the various specimens.

ing the later Paleozoic. (See fig. 4.) Another group, that of the echinoids (sea urchins), was present from the Ordovician on, but not particularly abundant. Toward the end of the Paleozoic the cystoids and blastoids declined to extinction. The crinoids declined greatly but survived. They continue to the present day and had a later, more modest expansion of a distinctive type (Articulata), but they have never regained anything like their early importance. The echinoids, previously a fairly common but minor group, also declined. Only a single type of echinoid (that of the cidaroids) pulled through from Permian into Triassic, but its later fate was quite different from that of the likewise reduced crinoids. From it during the Mesozoic there arose a great variety of new types and echinoids became far more abundant and varied than ever before. Thus the late Paleozoic expansion of the echinoderms involved mainly crinoids, with only slight contribution from other groups, and the Cretaceous-Tertiary expansion involved mainly echinoids. *Within* each of these groups there is a replacement of types. As between crinoids and echinoids there was succession but not distinctly replacement, because the two have quite distinct ways of life.

An example even more striking of this sort of evolutionary phenomenon is provided by the ammonites, a large group of extinct molluscs most nearly related to the chambered nautilus among living forms. These animals were very abundant during the Triassic, with twelve different families (according to a conservative tabulation prepared by O. Haas). In the Jurassic there were twenty-three families (same tabulation). But this is not, as the figures alone might suggest, a case merely of expansion as would appear normal at some phase in the history of a successful group of animals. It represents an almost complete replacement. Twenty-two of the twenty-three Jurassic families arose, and obviously arose with astonishing rapidity, from a single family that survived from the Triassic (Phylloceratidae); some students believe that all arose from a single genus within that family. The other eleven Triassic families became extinct. Thus the ammonites as a whole increased, but they did so by replacement, not by uniform expansion of the group as a whole.

Within the broader fabric of the evolution of phyla lie many details of this sort—obviously few such details can be followed within the scope of this book but others will be briefly examined later, among the chordates, by way of examples. It is a remarkable fact that in spite of the many extinctions within the phyla, shifting in proportions of their various groups, and replacements of some by others, the phyla as a whole did not become extinct.[5]

5. The extinct graptolites are often listed as a phylum and are given that rank on previous pages here for convenience. It is, however, seriously questionable whether the graptolites really represent a major, distinctive type of organization like the well-established phyla of the fossil record. They are themselves more uniform in type than members of most other

This is one of the most important of generalizations derived from the broad picture of animal history, and it has far-reaching implications. Before exploring these, however, it is advisable to lay a firmer basis for this and other generalizations by considering examples of more important details of patterns of the duration and diversity of various groups through time. Such examples can be obtained abundantly from any of the phyla with good fossil records; and the general sorts of events, as well as the evolutionary principles that they suggest, seem to be the same for all the phyla. Our examples will be drawn from the chordates, because their record is excellent, has been well studied, and is most familiar to me, as well as because it has an inherently stronger interest for man, who is a chordate.

phyla (excepting a few very minor recent supposed phyla of almost equally dubious status) and may simply be a peculiar subtype arising within some other phylum. The possibilities that this was the situation are strong but diverse. Students sometimes list the graptolites as a distinct phylum because they do not surely know to which (other) phylum they belong. A few other, still less well-known extinct animals (such as the Archaeocyatha) have been hailed as representatives of distinct and extinct phyla, but with very doubtful propriety. In spite of possible exceptions involved in the largely verbal question of defining "phylum," it remains true that *no major basic type of animal organization is known ever to have become extinct.*

IV. THE HISTORY OF THE VERTEBRATES

There are a few living chordates in which the dorsal stiffening rod (notochord) is not jointed: the acorn worms, sea squirts, and amphioxus, favorite of college zoology laboratories. These interesting groups cast some light on the origin of the phylum Chordata, but they have no real importance in the over-all economy of nature and (with dubious possible exceptions) they have no fossil record. They seem to represent, in modified and probably degenerate form, a stage rather rapidly traversed by the rising chordates in the early Paleozoic. Already in the Ordovician there had developed forms with more extensive skeletons and with the dorsal rod becoming modified into a jointed backbone. These forms, the vertebrates, have ever since been the dominant chordates, and it is their instructive history that is well recorded in the rocks.

The broad outlines of this record are shown in figure 5. Eight major types of organization are recognized, formalized as classes in technical classification. Four of these characteristically have fins and are primarily adapted to swimming in the water, while four characteristically have legs and are primarily adapted to walking on a solid surface. It is a typical complexity of evolution that these characteristic fins or legs may be lost in some modified lines and the primary adaptation profoundly altered or even reversed. Yet these exceptions among the protean lines of evolution need not confuse the simplicity of the over-all picture of four major grades of primarily aquatic and four of primarily terrestrial animals.

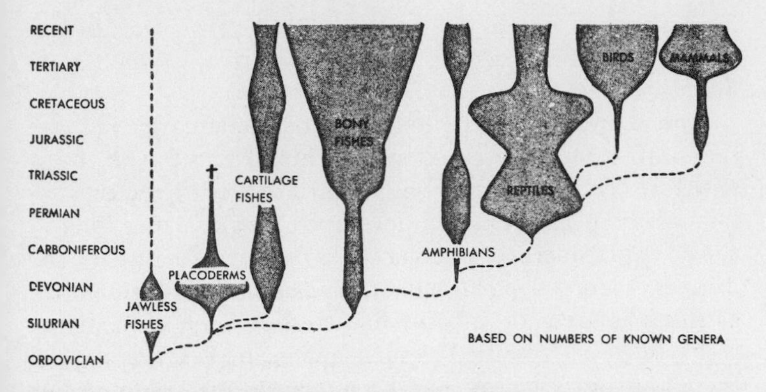

Fig. 5. The broad outlines of the historical record of the vertebrates. For each vertebrate class the width of the pathway is approximately proportional to its known variety in each of the geological periods in which it lived.

We are accustomed to thinking of all the primarily aquatic types as "fishes," but they represent at least four markedly different groups rather more distinct in fundamental anatomy than are reptiles and mammals. There are, first, the jawless "fishes," Agnatha, exemplified today by the lampreys. Their most striking peculiarity is that they lack a separately movable lower jaw jointed to the skull, so obvious a feature of all the other vertebrates be they sharks, trout, frogs, lizards, sparrows, or men.

Least familiar of the aquatic classes, because long extinct, is the Placodermi. Highly diverse in their day, these forms all had true movable jaws jointed to the skull in a somewhat primitive way. In the next class, the cartilage fishes or Chondrichthyes, evolution of the jaws becomes well advanced. The outstanding peculiarity of these animals is that bone, present in the placoderms and in the ancestors of the cartilage fishes becomes lost and the skeleton comes to consist of cartilage only, a substance somewhat weaker than bone although tough and resilient. Sharks, rays, and the less familiar chimaeras are living representatives of the Chondrichthyes.

Most successful, in the long run, of the aquatic classes is that of the Osteichthyes. They have an advanced type of jaw structure and, in contrast with the Chondrichthyes, they retained a bony skeleton and developed it still farther than in the bony placoderms.[1] Among other peculiarities of the bony fishes is the surprising fact that the possession of lungs, as well as gills, was one of their basic characters. In a few of the living forms lungs are retained as such but in the vast majority of them the former lung has been transformed into an air bladder and no longer serves for breathing. Almost all the familiar fishes of today are bony fishes: trout, catfishes, herrings, eels, perch, and a whole host of others—in fact, practically any fish you can name except the relatively few lampreys, sharks, and their relatives.

In the record the four aquatic classes appear in a sequence that seems to be that of their real origins: first the Agnatha in the Ordovician, then the Placodermi in the Silurian, and finally the Chondrichthyes and Osteichthyes, at about the same time, in the Devonian. Earlier students of evolution believed that the Chondrichthyes are more primitive than the Osteichthyes, perhaps ancestral to them, and should appear earlier in the record. With greatly increased knowledge of the record it now seems reasonably certain that neither is basically more primitive than the other, although the Osteichthyes have undergone more extensive evolutionary change since their origin. The two classes probably arose

1. By now it should come as no great shock to learn that this statement is true only over-all, as a generalization, and not in every detail. Evolution seems to have tried out almost every conceivable possibility and never to have followed a simple and uniform pattern. There are groups of "bony fishes" in which the trend has been to reduce bone and to have skeletons composed mainly of cartilage.

independently at nearly the same time and from rather similar ancestors as two different evolutionary solutions of problems of aquatic life. A reason for the appearance of two such distinct solutions in the Devonian is suggested by the possibility that the Chondrichthyes arose in marine waters and the Osteichthyes in fresh waters. A few Chondrichthyes later invaded the fresh waters and the generally more successful Osteichthyes early reached the sea and eventually became the dominant fishes there, as in the lakes and streams; but the original differentiation of the two classes may have corresponded with the two major divisions of aquatic habitats.

Each of the four aquatic classes first appears in the record with rather rare and little-varied representatives and each builds up within a period or two, that is, within some tens of millions of years, to a maximum. Aside from this their histories are strikingly different. After a climax in the late Silurian and early Devonian, the Agnatha declined almost to extinction, and yet a few peculiar types survived as relics through the millions of years to the present time.[2] The Placodermi had a somewhat similarly simple rise and fall, but they declined to the zero point, becoming extinct. In the Devonian they were the most abundant and varied of aquatic vertebrates. In the Mississippian they had already ceded this dominance to the somewhat similarly but better adapted Chondrichthyes and Osteichthyes, both of which probably arose from early placoderms—another example of replacement among major groups of animals. Placoderms lingered on, greatly reduced, into the Permian and then became extinct. It is probably not a coincidence that this extinction

2. They were rare, probably peculiar in habitat, and the surviving lines were characterized by complete loss of bone or other hard tissues apt for fossilization. As a result there is no fossil record of them during some 350,000,000 years, from Devonian to Recent, when they were nevertheless in existence. No such serious gap occurs in the records of abundant groups or of those with better development of skeletal tissues.

falls in the time of crisis for marine invertebrates mentioned above.

The Placodermi are the only one of the eight vertebrate classes to have become extinct. Major grades of organization in the phylum Chordata, like those still broader grades represented by the various phyla themselves, seem to be nearly, although not completely, immune from extinction.

The Chondrichthyes had already reached a maximum in the Mississippian and they declined thereafter to a low point in the Triassic, suggesting again that they shared in the Permo-Triassic crisis for marine life. But unlike the Placodermi they survived the crisis and came back to a new climax in the Cretaceous—there was (we might almost now say "of course") replacement of types within the class between its two maxima. The cartilage fishes have declined somewhat since the Cretaceous but still retain considerable absolute importance. In relative importance they have become greatly overshadowed by the bony fishes.

The bony fishes seem to have reached their first climax with unusual rapidity. They were already an important although not the dominant faunal element in the Devonian, soon after their origin. They then seem to have declined somewhat, but not radically, through the Permian. In the Triassic they were more abundant than ever before and took over the dominance of aquatic vertebrate life that they have held increasingly ever since that time. A moderate setback in the Jurassic was marked by the waning of some of the older types (especially the disappearance of a large structural grade, Subholostei, which was especially exuberant in the Triassic) and their incipient replacement. The main replacing group (called Teleostei in formal classifications but apparently a structural grade evolved from various Subholostei) arose around the beginning of the Jurassic and expanded at geometrically increasing rates thereafter. This group is responsible for the great preponderance of bony fishes in our streams and seas and it includes the majority of the 23,000 or so species of recent fishes.

The history of the four classes of walking vertebrates has some interesting parallels with that of the four classes of swimming vertebrates. In each case there are two successive grades of early forms, the second of which soon overshadows or largely replaces the first: Agnatha and Placodermi among swimmers, Amphibia and Reptilia among walkers. In each case, too, the second of these early classes gives origin, more or less simultaneously but independently, to two others, which rise jointly to dominance and which include the great bulk of the Tertiary and Recent faunas: Chondrichthyes and Osteichthyes among the aquatic, Aves (birds) and Mammalia among the land forms.

The parallel is worth stressing because in the tremendous multiplicity of nature any apparent regularity may be a clue to some essential principle of evolution. The comparison must not, however, be pushed too far. There are also radical and instructive differences between the fates of the four aquatic and of the four mainly terrestrial vertebrate classes. Among the aquatic classes the two earliest dwindled to insignificance and to extinction, respectively. The Amphibia and Reptilia dwindled, also, but each found special niches where there was no serious competition from other classes and in which they retained considerable importance. The Chondrichthyes and Osteichthyes, although different in original habitat, were to some extent parallel and came into competition at various points so that the success of one (Osteichthyes) has probably had an inhibiting influence on the other. The birds and mammals, on the other hand, arose with basically different adaptive characters and as a rule retained them, so that they have not seriously competed. (There are grounds for quibbling on the basis of some exceptions, but again we are discussing the picture as a whole and not worrying unduly over the endless complications in the included details.) Neither inhibited the other and both expanded greatly at about the same time.

The order of appearance in the record of these classes, as well as of the aquatic classes, is undoubtedly the order of their real origin and a consequence of their derivation one from another. Placodermi (appearing in the Silurian) arose from Agnatha (Ordovician), and both Chondrichthyes and Osteichthyes (Devonian) from Placodermi. Amphibia (late Devonian) arose from Osteichthyes (early Devonian), Reptilia (early Carboniferous) from Amphibia, and both Mammalia (late Triassic or early Jurassic) and Aves (Jurassic) from Reptilia. The earliest transitions, Agnatha-Placodermi, Placodermi-Chondrichthyes, and Placodermi-Osteichthyes are not well recorded (as the record is now known) but are well-supported inferences from the available evidence. The transitions Osteichthyes-Amphibia, Amphibia-Reptilia, Reptilia-Mammalia, and Reptilia-Aves are all clearly shown in the record. There are, of course, missing minor steps and more evidence as to details will be welcome, but the material in hand already leaves no doubt as to the reality and the essential features of these derivations of one class from another.

The four terrestrial classes agree with the four aquatic in having slender beginnings and rising with greater or less rapidity to a maximum. The amphibians are moderate in this respect, arising in the late Devonian and perhaps reaching their high in the Carboniferous, later by a span on the order of 60,000,000 years. The Amphibia continued little below this high level through the Triassic, but from the Permian onward they are over-shadowed by the far more numerous and varied Reptilia. After the Triassic the amphibians declined radically, nearly to extinction but not quite, because two main lines, then obscure but evolving new sorts of adaptations, pulled through: that of the frogs and their allies (the order Anura) and that of the salamanders

(Urodela).[3] These groups, and particularly the former, were involved in a second, more modest, expansion of the Amphibia during the Tertiary and they have won for the class a rather small but significant place in the present life of the continents.

Superficially, this history of the Amphibia suggests replacement within the class, analogous to the replacement of earlier types by the Teleostei among the Osteichthyes, or to the replacement of Triassic ammonite families by an array of new groups derived from one only of these families. There are, however, essential differences in the evolutionary phenomenon exemplified by amphibian decline and reexpansion. In the examples of the bony fishes and the ammonites the later forms are, on the whole, essentially similar to those they replace. By and large, they led somewhat the same sorts of lives and occupied (or, at least, included in their wider range) much the same places in the economy of nature. They literally replaced the earlier groups in their own sphere. The later amphibians did not, in this sense, replace the earlier. The salamanders clung, more or less precariously, to only one minor part of the earlier range of amphibian adaptations. The more particularly successful frogs developed new and peculiar adaptations which involved ways of life not, as far as we know, open to previous amphibians, or to any earlier animals. The most abundant types of earlier amphibians were replaced not by later amphibians but by reptiles —and perhaps to some extent also by fishes of the newly successful Mesozoic groups.

Early reptilian expansion was extraordinarily rapid. From slim beginnings in the Carboniferous, the reptiles were al-

3. There is another living order of amphibians, the Apoda or Gymnophiona, including a few, obscure, superficially wormlike animals, but they are of no great importance. Their fossil history is poorly recorded.

ready highly varied and abundant in the Permian.[4] In spite of the usual textbook tags of "Age of Amphibians" for the late Paleozoic and "Age of Reptiles" for the Mesozoic, reptiles were at least as dominant among vertebrates and among forms of land life in the Permian as in the Triassic.[5] Reptiles continued abundant throughout the Mesozoic, with perhaps some slight relative diminution in the earlier Jurassic. This, if it is not merely a chance effect of the record, reflects extinction or diminution of several Triassic and earlier orders before others, replacing them or extending reptilian occupation into new habitats, had risen to their great Cretaceous climax. Reptiles were actually most varied in fundamental characteristics in the Triassic, but the Cretaceous apparently was their greatest period in absolute abundance and in variety of lesser types.

Latest Cretaceous and earliest Tertiary times saw the great crisis in reptilian history. Like the Jurassic crisis in amphibian history, this was a phenomenon of restriction without replacement from within the class. Four major groups (orders) out of ten present in the Cretaceous survived this crisis. All of these declined from their earlier high points, but three of them, those of the turtles (Chelonia), the snakes and lizards (Squamata), and the crocodiles (Crocodilia), retained considerable importance through the Tertiary and

4. This rapid expansion is clearly real, but it is somewhat exaggerated by the data for the Therapsida (mammal-like reptiles) incorporated in figures 5 and 6. The counts of genera for this group in the Permian are inflated relative to those other fossil reptiles and other periods by the intensity of study of South African Permian reptiles and the tendency of their leading students to base a multitude of genera on relatively minor variations in structure. The same sort of artificial fluctuation certainly affects other groups as well, notably among the invertebrates.

5. The usual "Age of Mammals" for the Cenozoic is more apt, although "Age of Teleost Fishes" and "Age of Birds" may be equally defensible.

into the Recent. The other surviving order, Rhynchocephalia, never a very large or dominant group, was reduced almost to extinction and survives only in a single relict form on islands off New Zealand, the lizardlike *Sphenodon* or *Hatteria.*

This creature, it may be noted in passing, represents one of the most remarkable examples of evolutionary stagnation. It is almost identical in structure with ancestral forms from the Jurassic, and the evolutionary rate of this line has been virtually nil for the last 140,000,000 years or more. Restriction of this line to the New Zealand region or to areas comparably secluded and paleontologically poorly known may date from the Cretaceous, for no fossils are known after the early Cretaceous. Disappearance of sphenodonts from the record occurs around the time when the adaptively similar lizards were beginning their expansion. It appears that the sphenodonts were replaced by lizards except in a remote asylum where lizards probably did not penetrate.

The bird and mammal records are comparable, and both are peculiar, in that after their origin both groups seem to have remained obscure and rare for a long time before exhibiting any considerable expansion. Both groups appear in the Jurassic (mammals doubtfully in the latest Triassic), but the record for both is scanty until the early Tertiary, more than 100,000,000 years after they originated. Birds and mammals were by no means stagnating during this long period of their obscurity. There is evidence that in each case the most fundamental parts of their structural evolution and basic divergence occurred during this time. It is, however, most unlikely that they were important and widespread in the faunas of the world as a whole. They seem to have been evolving in rather small local groups or in restricted habitats that happen not to be continuously represented in the fossil record.

Birds and mammals both expanded remarkably in the

early Tertiary, and both tended to replace Cretaceous groups of reptiles, although in addition to partial and approximate replacement both also came, in their diversity, to embrace ways of life followed by no earlier animals. To the extent that replacement was involved, this example serves to illustrate the possibility of confusion when the record is interpreted in terms of causes. The record shows abundant reptiles and few birds or mammals in the late Cretaceous, then few reptiles, more birds than before (but still not very many), and abundant mammals in the early Tertiary. There is no logical support here for the conclusion that the reptiles dwindled, and many became extinct, *because* they were replaced by birds and mammals. The facts are better fitted by (and yet they do not prove) the reverse proposition, that birds and mammals replaced reptiles because the reptiles had dwindled or become extinct.

The record for birds is none too good—small, thin-boned, flying animals do not fossilize very readily—but it suggests rapid basic expansion in the early Tertiary, into the Miocene, followed by increasing minor diversity within the groups thus established, down to the present day with its total of about 8,600 bird species. Mammals reached their high point in the Pliocene and have since declined somewhat to a total of species estimated at 4,500.

V. RECORDS AND RADIATIONS OF REPTILES

To adduce examples of life histories on a smaller but still rather broad scale, the record will be briefly reviewed for the various orders included in the two classes of animals nearest to us, that of the mammals, to which we belong, and that of the reptiles, from which mammals arose.

The known record for the Reptilia is diagrammatically summarized in figure 6.[1] It is unnecessary here to review each of the sixteen orders into which the class is divided, but attention is called first of all to the diversity of their patterns of (known) distribution. Few show what might, a priori, be considered the most likely history for a group of animals: origin, gradual rise to a climax, and gradual decline to ex-

1. The orders listed are those so recognized in A. S. Romer, *Vertebrate Paleontology* (2d ed. Chicago, University of Chicago Press, 1945). The numerical data for the Reptilia were also derived mainly from Romer, as previously acknowledged. Numerous fossil reptiles have been discovered in the last 20 years or so, but that does not significantly change the relative diversity of the various orders. The most evident change brought about by recent studies is that the "Ictidosauria" of the figure are not now to be considered a united and major group of reptiles. The group so labeled in the figure includes various separate groups of mammal-like reptiles (therapsids or therapsid-derived) close to or, in one or two instances, perhaps over the arbitrary line taken as dividing mammals from their nominally reptilian ancestors. Note also that the exigencies of space make the figure somewhat misleading as to descent and relationships of the various orders. Notably, the Saurishia and Ornithischia (the two orders of dinosaurs) had a common origin in the Thecodontia. The Pelycosauria had a different origin and were ancestral to the Therapsida.

tinction. This is the pattern for the two oldest, most primitive orders, Cotylosauria and Pelycosauria, but it is not clearly seen in the record of any other order. Several reach a maximum near, or practically at, the time of their first recorded occurrence and then decline slowly (Eosuchia), moderately (Ichthyosauria, Rhynchocephalia, Pterosauria), or relatively

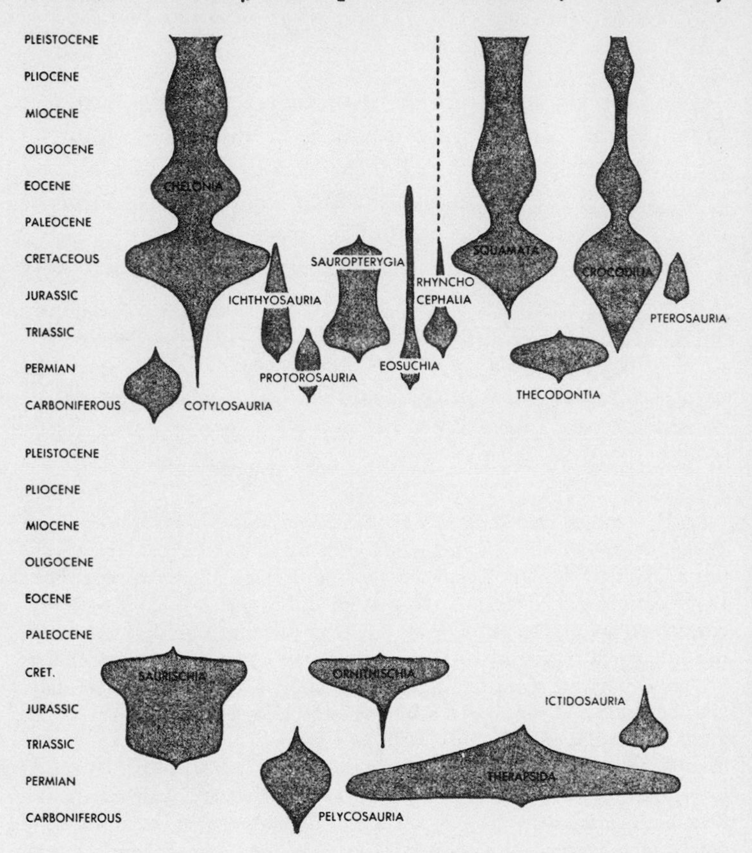

Fig. 6. The historical record of the reptiles. For each recognized order of reptiles the width of the pathway is proportional to its known variety in each of the geological periods in which it lived.

rapidly (Therapsida) to extinction. One only, the dinosaurian order Ornithischia, expands rather slowly to a climax almost immediately followed by extinction. Several, particularly the plesiosaurs, Sauropterygia, and the other dinosaurian order, Saurischia, are already highly varied and abundant at or soon after their first appearance and continue at a high level until their geologically speaking abrupt disappearance.

It is noteworthy that the three living orders (excluding the interesting but insignificant living Rhynchocephalia) have closely similar patterns on the record. They originate rather obscurely, expand in a gradual way to strongly marked maxima in the Cretaceous, and then continue with fluctuations to the present day. The record for all of them shows an abrupt restriction in the Paleocene followed by a new expansion, less than for the Cretaceous, in the Eocene, and then by slight reduction to the Recent, but this is probably an artifact of the record and not a phenomenon of the groups as they really existed. Paleocene reptiles happen to be poorly known and it is certain that for that epoch the number of genera known is a considerably smaller proportion of those that existed than for the Cretaceous or the Eocene. It must also be remembered that the Cretaceous was probably at least three times as long as the Eocene, or any other Tertiary epoch, and therefore might be expected to have more total known genera even though no more were then living at any one time.[2] When allowance is made for these factors, the data suggest

2. The varying lengths of the periods involved have undoubtedly similarly influenced the other patterns of distribution discussed here. It is, however, unlikely in other cases that this effect has seriously falsified such general and relative features of the patterns as have been mentioned in the text. Anyone who wishes to go beyond me in the interpretation of details of these patterns should beware of this and of several other sources of error and uncertainty, which I have discounted or avoided in this discussion but which might be misleading in any more detailed comparisons.

that the turtles (Chelonia) and the snakes and lizards together (Squamata) have retained about the same abundance and variety from Cretaceous to Recent times, while the Crocodilia have declined somewhat but probably not as much as would appear in the diagram and without so strong a Paleocene recession.

Another sort of evolutionary phenomenon, which could indeed be exemplified by many other groups of animals, is particularly well shown by the reptiles: adaptive radiation.[3] Hitherto in this discussion the stress has been on the fact that there are major structural and functional grades and types, which have run their courses and waxed and waned in the course of geologic time as their various potentialities arose, were developed in different lines of descent, or were taken over by one group from another. Now it is time to exemplify the remarkable variety of such potentialities even within a group arising from one basic structural type, to suggest the way in which these potentialities can expand in the course of evolution and can enable their possessors to push into environments and to follow ways of life radically different within the group and radically different from those of its ancestral and basic type. Adaptive radiation is, descriptively, this often extreme diversification of a group as it evolves in all the different directions permitted by its own potentialities and by the environments it encounters.

The basic type of reptile is a vertebrate living on land throughout its life, including the time when it is developing as an embryo, laying eggs and taking little or no care of its

3. The term was particularly employed and I think was coined by the late Professor H. F. Osborn. The phenomenon so labeled was, however, well known to Darwin and was clearly described in other words in *The Origin of Species*. It is discussed in most general works on evolution, for example in G. G. Simpson, *The Major Features of Evolution* (New York, Columbia Univ. Press, 1953) and V. Grant, *The Origin of Adaptations* (New York, Columbia Univ. Press, 1963).

young, equipped with four stout legs on which it can walk or run rather clumsily, long and low in body, with a large number of little differentiated conical or peglike teeth, and with a complex brain adequate to control this bodily apparatus but relatively poor in the higher (or, at least, later developed) associative centers of the cerebrum. Reptiles answering to this general description are the first to appear in the record, among the cotylosaurs and pelycosaurs of the late Carboniferous and early Permian. Later forms lost or greatly modified almost every one of these features of the basic reptile, although reptilian in origin and still reptilian by definition. (See fig. 7.)

The two most radical departures were perhaps the development of aquatic and of aerial reptiles. In fact, several different aquatic types developed independently. Most extreme were the ichthyosaurs (order Ichthyosauria). They developed a fishlike tail and a dorsal fin. Their legs were reduced and modified to paddles, steering instruments quite incapable of supporting the body. Thoroughly aquatic, they became amazingly fishlike in external form and passed their whole lives in the sea. Unable to come to land to lay eggs (as marine turtles do), they became ovoviviparous, that is, the eggs were incubated within the body of the mother and the young appeared in the outer world alive and kicking, or rather swimming. (Rattlesnakes and a few other present-day reptiles are also ovoviviparous, but mostly in adaptation to different exigencies.) Yet they continued to breathe air and every bone in their bodies was different from that of any fish, although adapted to thoroughly fishlike activities.

Another thoroughly aquatic but less fishlike type of reptile is seen in the plesiosaurs, members of the order Sauropterygia. They had large, rather flattened bodies, more or less turtlelike though usually not armored. The legs became large flippers with which the animals sculled their way along. The tail was tapering and the neck was often extraordinarily long.

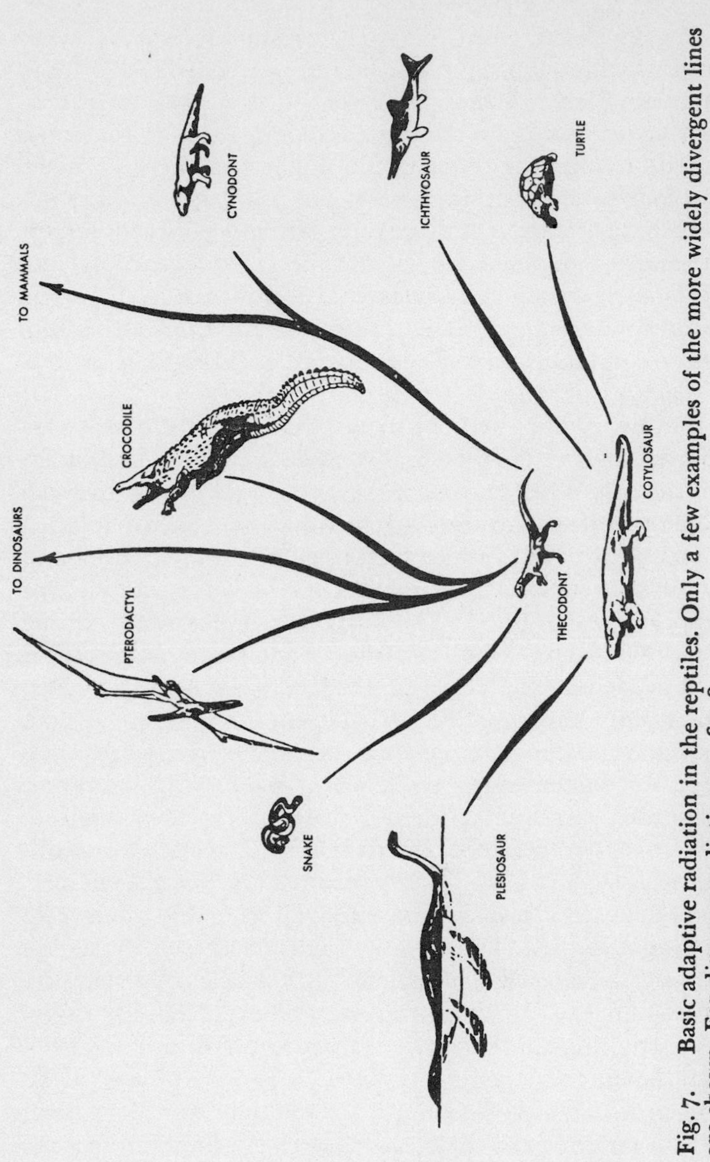

Fig. 7. Basic adaptive radiation in the reptiles. Only a few examples of the more widely divergent lines are shown. For dinosaur radiation see fig. 8.

Besides these two whole orders of aquatic reptiles, many turtles are aquatic, as is well known; a group of lizards (the mosasaurs) became thoroughly aquatic in the Cretaceous—real sea serpents; and the crocodiles as a group are basically amphibious and some of them became completely aquatic. Various other amphibious, or even thoroughly aquatic, lesser types have developed from time to time among diverse groups of reptiles, including the snakes, lizards, and dinosaurs. (No dinosaurs are known that were confined to life in the water, but many were amphibious.) In fact, invasion, or re-invasion, of the domain of their piscine and amphibian forebears was a widespread characteristic of reptilian history.

Flying reptiles form the order Pterosauria, a group birdlike in some respects, still more batlike, but with reptilian peculiarities of their own. They flew by means of a leathery membrane supported and extended by an enormously elongated finger; not by several long fingers on each hand as in the bats.

In the familiar lizards the general ancestral reptilian type is not very much modified except as these animals tend to be smaller and capable of more rapid locomotion. Yet even within this one group there has been much minor radiation, producing flattened, spiny forms, others that are bipedal, and even some that are limbless and snakelike. The highly specialized snake type is also familiar to all. It has less evident potentialities for secondary radiation, yet snakes have become highly varied in habitat, diet, and other respects.

The turtles (order Chelonia) have developed armor of bone and shell and have lost their teeth, replacing them with a horny, birdlike beak. These specializations would seem to restrict further opportunities for radiation within the group, yet turtles live under a great variety of conditions, from the high seas to stony deserts, and follow many diets, from strict vegetarianism to a regime as carnivorous as that of the tiger.

The general crocodilian type, quadrupedal, amphibious to

aquatic, plated down the back, and carnivorous—but of varied diets—exhibits an interesting example of replacement among the reptiles. The type arose first in the Triassic in a group of reptiles called phytosaurs (members of the order Thecodontia). The phytosaurs were completely replaced by the crocodiles, of different origin but of essentially the same adaptive type, in Jurassic and all later times.

A wide variety of opportunities for medium-sized to large land dwellers was exploited and pre-empted by the reptiles collectively called dinosaurs, belonging to the two orders Saurischia and Ornithischia.[4] (See fig. 8.) The dinosaurs within themselves underwent an adaptive radiation extraordinary in its diversity. End products, all derived from ancestors closely similar to each other and little removed from the basic reptilian type in the Triassic, included such creatures as:

Struthiomimus—a bipedal, slender, remotely ostrich-like, hollow-boned, swift-running dinosaur with no teeth.

Tyrannosaurus—also bipedal, but heavy, huge (some twenty feet high when erect), big-headed (skull over four feet long), strictly carnivorous and with rows of awe-inspiring, daggerlike teeth.

Brontosaurus[5]—the popular image of "*the* dinosaur," thanks to excellent publicity, a hulk of thirty tons in weight, more or less, and some seventy feet in length, with four post-like legs, a whiplash tail, and a long neck carrying a ridiculously small head for so big a beast, with teeth fitted only for gathering soft, aquatic vegetation as food.

4. There are many books on dinosaurs, most of them too childish or, as the French would say, too vulgarized for the readers of this book. Among the exceptions is E. H. Colbert, *Dinosaurs* (New York, Dutton, 1961).

5. The rules of zoological nomenclature would have us call this monster *Apatosaurus*, but, rules or no rules, the name *Brontosaurus* has become known to every schoolchild and is undoubtedly here to stay.

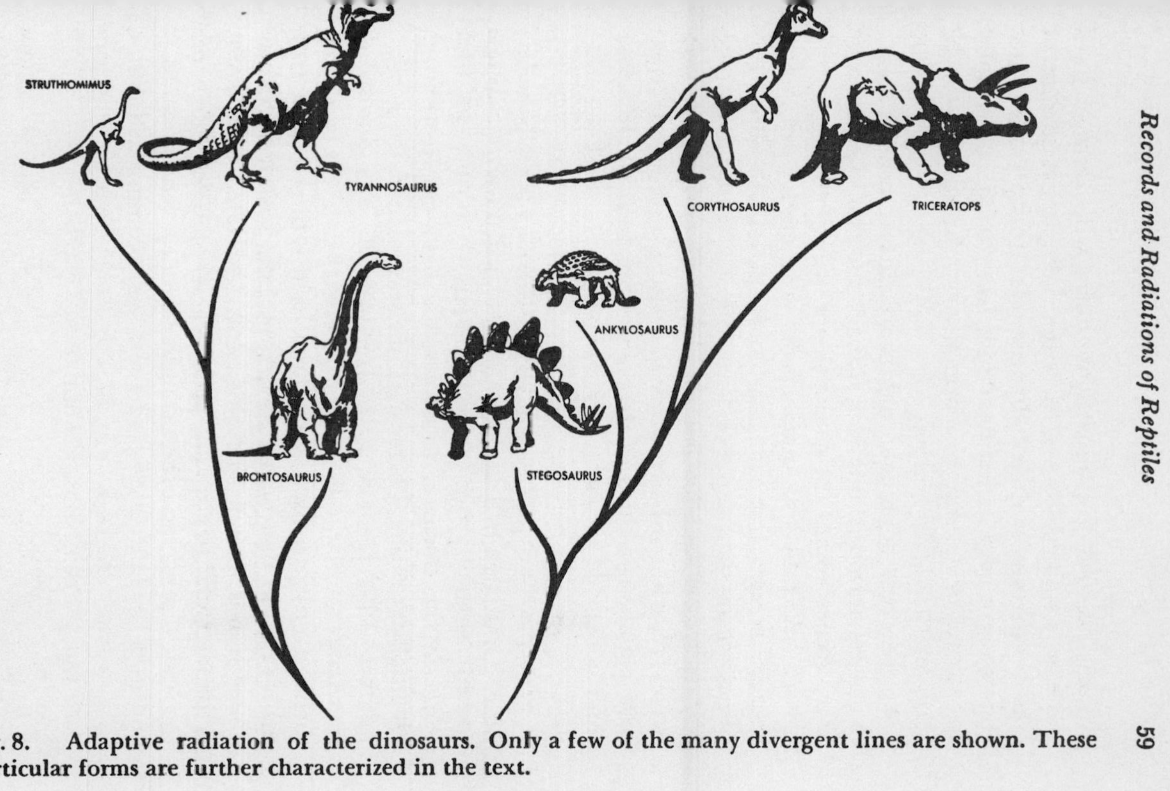

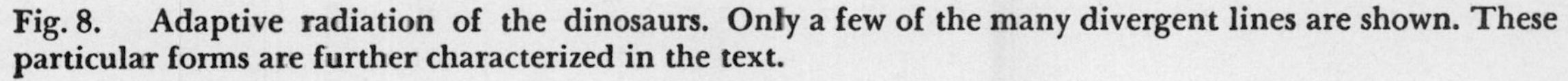

Fig. 8. Adaptive radiation of the dinosaurs. Only a few of the many divergent lines are shown. These particular forms are further characterized in the text.

Corythosaurus—an amphibious form with webbed feet, more or less bipedal out of the water, with an extraordinary crest on the top of the skull suggestive of a rooster's comb (but made of bone), a bill something like that of a duck, and, in the back of the mouth, a large battery of teeth adapted to chopping harsh vegetation.

Stegosaurus—a quadrupedal form with two rows of large, flat plates down the middle of its back and two pairs of long, sharp spines toward the end of the tail, also herbivorous.

Ankylosaurus—also quadrupedal and herbivorous, but encased in rigid armor of many bony plates and with long spines on the shoulders.

Triceratops—another but markedly different quadrupedal herbivore, without body armor but with a tremendous head covering about a third of its twenty-foot total length, with three horns, and with a bony frill extending far back over the neck.

These, and other types equally distinctive, are all the results of radiation within two closely related stocks of reptiles within a span of about 120,000,000 years.

Finally, mention must be made of the mammal-like reptiles, essentially the order Therapsida. These, too, included a large number of divergent subtypes, but as a unit in the broader reptilian radiation their special interest lies in the fact that their adaptive trends were mainly in the direction of the mammals and that mammals did, in fact, arise from them. These trends included developments correlated with more active and effective quadrupedal locomotion than that of the basic reptile and others concerning the eating apparatus and associated structures. The legs were coming to lie farther in under the body, raising it to a posture more efficient than the sprawling position of most early reptiles. The teeth were becoming less numerous than they had been, but more differentiated according to function in different parts of the jaws. The lower jaw, which in other reptiles consisted of several different bones of nearly equal size, was coming to be formed mostly by the one pair of large, main tooth-bearing bones. Associated changes were going on in the jaw hinge and in the ear, which lies near this hinge and is affected by changes in that region. Unfortunately there is no way of telling whether the advanced therapsids cared for their young and had milk to feed them, but this is not unlikely. There is some, but inconclusive, evidence that they were becoming warm blooded. It would be gratifying to our mammalian vanity to be able to add that the therapsids were becoming more intelligent than other reptiles. The fact is, however, that their brains were still thoroughly reptilian in grade, and so were those of the earliest mammals, as will be mentioned again when we come to discuss the mammals.

In the midst of all this radiation of the class Reptilia, the basic reptilian type came to be relatively rare and exceptional. Indeed, it nowhere survived in quite its original form. The expansion of the Reptilia was not, then, simply a matter of the increasing abundance of one general type of animal. It involved, rather, the development of many and strikingly different types from an ancestry that had reached a new grade of organization. Some of these types converged toward, competed with, and occasionally quite replaced other animals that had earlier developed in radiation from more ancient grades of organization. So the ichthyosaurs must have competed with similar types of sharks (which were, indeed, reduced in abundance during the span of the ichthyosaurs), and so, after the Triassic, amphibians similar in way of life to amphibious reptiles are no longer found. Many, evidently the majority, of the radiating types of reptiles, however, became forms such as the earth had never seen before and followed ways of life that were new under the sun.

The reduction of the Reptilia, too, was not a matter simply of the waning of a grade of organization, or even of general decimation of all the various types developed from such a grade. It meant the elimination, total and irrevocable, of most of the divergently specialized sorts of reptiles and the survival, with scarcely diminished success, of a few. The survivors were not, as a whole, notably more primitive or less remarkably specialized than those that became extinct. On the contrary, their structures and their ways of life were in many cases among the most peculiar and narrowly delimited of those developed by any reptiles. The survivors were types that had occupied niches in the ecology of the earth not successfully filled by any earlier forms and not, as it happened, closely approached by any of those that came later.

VI. OUTLINES OF THE HISTORY OF MAMMALS

The rise of the mammals involved the development of numerous interrelated anatomical and physiological characters that proved in the long run to be more effective in many (not all) of the spheres of life occupied by the reptiles. These also were, in the course of time, a basis for the development of new ways of life never achieved by reptiles or by other forms arising within earlier adaptive radiations. The evolution of these new and, as the outcome proved, potent features began among certain of the reptiles, and very early in reptilian history. In a sense the mammals, and the birds too, are simply glorified reptiles. But in a similar sense the reptiles are glorified amphibians, the amphibians glorified fishes, and so on back until all forms of life might be called glorified amebas,[1] and the very amebas might be considered glorified prebiological molecules. The point is that a particular sort of reptilian development turned out to have such unusual possibilities for diversification and for the rise of novel and successful types of organization that its outcome came to overshadow that of all other reptiles put together. The zoologists therefore label that outcome as a distinct Class Mammalia, on a level with the Class Reptilia of which it is a particularly flowery branch.

Among the many developments within this potent reptile-mammal line, care of the young must be given high place.

1. To use "ameba" in a merely figurative sense for the not specifically amebalike, archaic protozoan level of life where cellular organization had first been reached.

Eggs were no longer deposited and left at the mercy of an egg-hungry world, nor even given such lesser care as external (as in birds) or internal (as in some reptiles) incubation. The embryo developing from the egg was continuously nourished, by intricate and marvelous means, within the body of the mother.[2] After being introduced to the world the young still receive care from their mother and are nourished for a time by milk from her. These animals also came to be adapted for a higher or more sustained level of activity and for a more constant level of metabolism. Most of them maintain a body temperature relatively independent of temporary activity or of external heat and cold. (This is what is meant by being "warm blooded"; a "cold-blooded" animal may have warmer blood than a mammal if it has been exercising violently or is exposed to the hot sun, but its blood cools down again when it stops muscular action or when it moves out of the sun.) The bones of the skeleton grow in a way that maintains firm, bony joints even while they are growing. Growth ceases and the bones knit firmly at an approximate size characteristic for each kind of animal. These arrangements are mechanically stronger than in animals, like the typical reptiles, in which

2. Everyone knows that this is another of the vast majority of generalizations that are open to exception. The platypus and the echidna lay eggs and are called mammals, but viviparity as opposed to oviparity is nevertheless a mammalian characteristic. It is even likely that it had arisen already in mammalian ancestors that we call reptiles, and the egg-laying mammals may derive from some other, allied, line of (nominal) reptiles that did not happen to develop this particular mammalian character. It is well known, too, that in one group of mammals, the marsupials, protection and nourishment of the developing young are less perfected than in the great majority of mammals, the many placental groups—but the fact that an evolutionary development may occur in greater or less degree or under different forms does not make it less characteristic of a group as a whole or lessen the importance of its degree and form in the majority of the members of the group. Incidentally, the platypus and echidna do have milk and are not exceptions to the next statement.

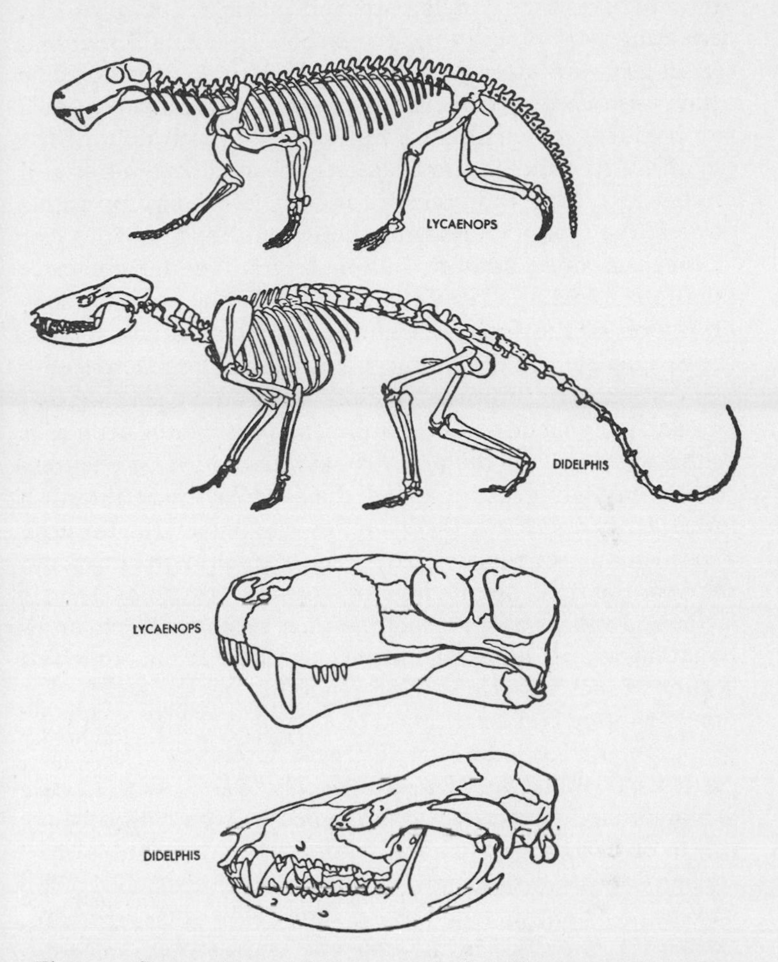

Fig. 9. Comparison of skeletons and of skulls of an extinct mammal-like reptile and of a primitive living mammal. (*Lycaenops* is from the Permian of South Africa; its skeleton was reconstructed under the direction of Dr. E. H. Colbert. *Didelphis* is the Virginia opossum. The drawings are to different scales: *Lycaenops* was considerably larger than *Didelphis*.)

the joints are more cartilaginous and continue to grow at decreasing rates through most or all of the animal's life. In connection with these features, the basic type of mammal, which was quadrupedal, came to have the legs drawn in directly under the body and to hold the body well up off the ground. This stance led to characteristic modifications of almost every muscle and bone in the body in comparison with those of the typically sprawling basic reptiles.

More sustained activity and more constant metabolism require considerable regularity of food intake and efficient utilization of the food. Evidently connected with this was the development in the reptile-mammal line of teeth specialized by regions: nipping incisors in front, then larger, pointed, piercing or tearing canines, and then a row of cheek teeth (premolars and molars), diversely fashioned for seizing, cutting, pounding, or grinding the food before it is swallowed. Early in the definitely mammalian part of the history a particularly important basic cheek tooth type was developed: the tribosphenic type, with several points or cusps, crests, and valleys on each tooth, so that seizing, cutting, and pounding can all be performed at once. Evolutionary modification of such a tooth, with emphasis of one part and function or another and duplication or extension of the pertinent parts, can and has led to the divergent development of teeth particularly suited for almost any conceivable diet. These dental developments were accompanied by direct jointing of the tooth-bearing jawbone to the skull and by increased strength of jaw action and versatility in directions of jaw movement. This change was, in turn, correlated with an extraordinary change in the ear. The single vibration-transmitting bone in the middle part of the reptilian ear was replaced by a chain of three small bones in the mammals and the two extra bones are parts of the old reptilian complex jaw joint. Other changes in the mouth region included development of a hard secondary palate between the mouth and the nose passages with the result that mammals can easily chew and breathe at the same time.

Many other changes were involved in the reptile-mammal transition, but enough have been noted to exemplify, in a general way, the sort of thing involved in the rise of a new grade of animal organization. Many of these changes were already under way among the therapsid reptiles of the Triassic. As far as can be shown by fossils, all had been essentially established in the Jurassic, in which five distinct sorts (orders) of mammals are known from unfortunately scanty remains. One of these groups, the rather rodentlike Multituberculata, continued with increasing importance through the Paleocene but became extinct in the Eocene—probably a case of replacement by the true rodents. All other later mammals probably arose from one Jurassic group, Pantotheria, in which the tribosphenic type of tooth, mentioned above, was being developed.[3]

The general outline of the known fossil record for mammals is shown in figures 10–13.[4] For some of the orders this knowledge probably corresponds reasonably well with the real abundance and variety of the animals through their history. In other cases it certainly does not. The correspondence, although necessarily only approximate at best, may be considered fair or adequate for most of the extinct Tertiary orders (except a couple of the comparatively rare and un-

3. The quaint living monotremes, platypus and echidna, of Australia probably did not have this origin, although some authorities think they did. Their origin is really quite unknown, but my own (not particularly original) suspicion is that they are rather highly modified surviving therapsid reptiles, mammals by definition rather than by ancestry.

4. In figure 5 the data for mammals are little modified from Romer to keep them comparable with those for the other vertebrates. In figures 10–13 and all the graphs dealing with mammals alone the data are taken with later additions and corrections from G. G. Simpson, "The Principles of Classification and a Classification of Mammals" (*Bull. Amer. Mus. Nat. Hist.*, *85* [1945], i–xvi, 1–350).

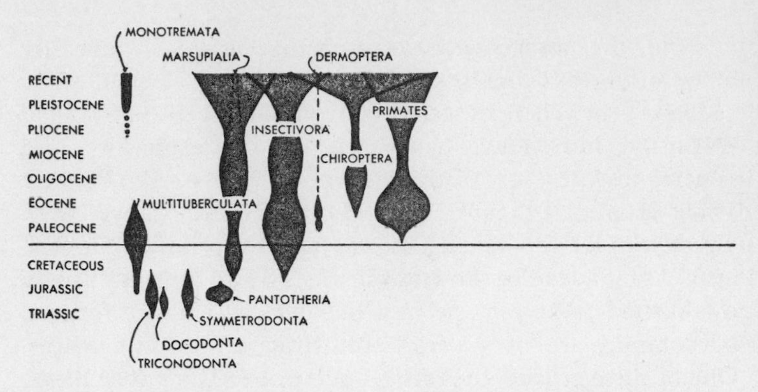

Fig. 10. Part of the historical record of the mammals. Each order of mammals is represented by a pathway the width of which is proportional to its known variety during the various periods and epochs in which it occurs. The record is continued, with the same scale, in figs. 11–13.

important ones). It is probably also reasonably good for the armadillos, sloths, anteaters and their allies (order Edentata), the rabbits and their kin (Lagomorpha), the carnivores (Carnivora),[5] the elephants and related forms (Proboscidea) and for the two great ungulate orders Perissodactyla and Artiodactyla. The marsupial record can be considered good if Australia is left out of question and the diagram is followed from the Pliocene on by the broken and not the solid lines. Pre-Pleistocene marsupials in Australia are poorly known, although they certainly existed in great numbers. The paucity of this knowledge and the great numbers of Pleistocene and Recent Australian marsupials falsify the diagram as a whole.

The form of the insectivore record (represented today by the shrews, moles, hedgehogs, etc.) shows an early Tertiary maximum and subsequent slow contraction. The early maximum and also the Cretaceous occurrences represent a number

5. Except that the apparent expansion from Pleistocene to Recent is probably not real.

of extremely primitive placental mammals, most of them not ancestral to the relatively specialized now living insectivores and some of them close to or even in the ancestry of some other orders of mammals. Later nominal insectivore records, from about the Miocene onward, do represent Insectivora in the special sense, as represented by shrews, moles, hedgehogs, tenrecs, and few other groups. The apparent sharp expansion in the Recent is largely or wholly an artifact, because we know all the living forms but only a fraction of their small, obscure, fossil ancestors and relatives.

The record for bats (Chiroptera) is so bad as to be almost meaningless in terms of relative abundance and variety, although it has great interest in some other respects. Bats simply do not get preserved as fossils except by rare accident and they certainly have been many—but an unknown number of—times more varied throughout the Tertiary and Pleistocene than is shown by the record.

The primate record can be qualified as about halfway good, or bad. Because of its special interest, it will later be the subject of separate discussion and more detailed evaluation.

The records of the scaly anteaters (Pholidota) and farther along, of the aardvarks (Tubulidentata) are both wholly inadequate; but for present purposes this does not particularly

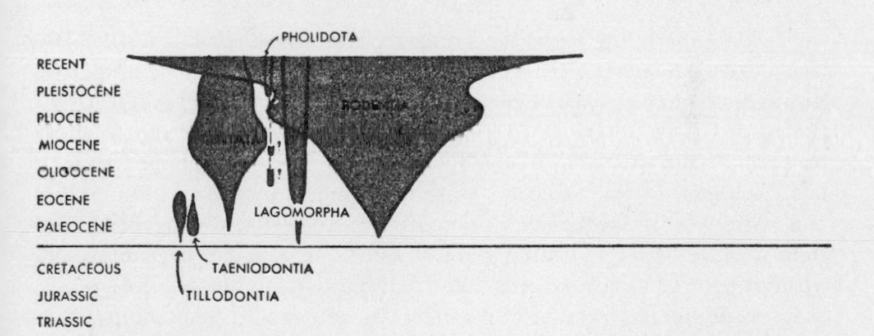

Fig. 11. Continuation of the historical record of the mammals from fig. 10.

matter because these animals are not and pretty surely never have been of any real numerical importance in the mammalian faunas of the world. Much the same can be said of the coneys or hyraxes (Hyracoidea) and the sea cows (Sirenia), although their importance is slightly greater and their records are fair but not really adequate.

The rodent record seems to me to be good, although, like that of the Insectivora, it requires one rather obvious correction if it is to correspond roughly with the facts. We know nearly all the living rodents but probably fewer than a third of the ancient rodents for each epoch. The distribution form would therefore probably be reasonably true to life if the record of known groups (genera) for the Recent were reduced by a third or those for each earlier epoch multiplied by three.[6]

Finally, in general form the known record for the whales, dolphins, and their allies (Cetacea) may correspond rather well with their real history except that there has probably been a fairly steady reduction since the Miocene and no expansion from Pleistocene to Recent.

After making these corrections and reservations, it is seen that the mammalian orders have much the same variety of histories of expansion and contraction as do the reptilian orders. Some build up to a high level, maintained for a longer

6. Some specialists on rodents disagree that knowledge of fossil forms is as nearly adequate as this, and suggest that rodents were as numerous throughout most of the Tertiary as they are now. However, I feel that the record does not warrant this conclusion and that it makes more probable the inference that not until the Pliocene, or perhaps but with less likelihood in the Miocene, were rodents comparable in abundance and variety with the present time. In the Eocene and Oligocene they seem to have been expanding rapidly, but from a scanty beginning and without yet reaching anything like their present dominance. Moreover, as far as numbers of species and genera are concerned, this dominance is strongly increased by the family Muridae, which seems to have been of relatively late origin.

or shorter time, and then taper off slowly to extinction (Litopterna, Notoungulata) or near it (Proboscidea). Some reach a maximum soon after they first appear and then decline rapidly (Condylarthra) or slowly (Perissodactyla). None expands slowly to a climax and then rapidly disappears. Only the Ornithischia did this among reptiles and the pattern seems to be rare among animals in general. The rodents seem to demonstrate a pattern which might be expected as the usual one but which is, in fact, exceptional: fairly steady increase throughout their history. It is, indeed, possible that this is true of rodents only up to the Pliocene and that they have since declined.

Carnivores and artiodactyls, two of the dominant orders from early Tertiary times to now, are similar in showing quick rise to abundance soon after their first appearance and in maintaining a high level ever since. Both also show a second expansion to still higher level in the later Tertiary, the carnivores in the Miocene (maintained into the Pliocene) and the artiodactyls in the Pliocene (beginning but less noticeable in the Miocene). The carnivores have probably declined since the Pliocene and the artiodactyls have surely done so, although both are still among the most varied of orders.

Some of the broader details of internal structure in these two groups (also shown in figs. 12–13) are of particular interest as examples of the sort of phenomena involved within the histories of the orders. The early, rapid expansion of the carnivores was almost entirely due to archaic forms known as creodonts, which formed almost the whole of the Paleocene and still decidedly the greater part of the Eocene carnivore population of the earth. During this Paleocene and Eocene dominance of the creodonts, they underwent a well-marked and quite diverse adaptive radiation. Beginning modestly in the Paleocene, a new basic carnivore type arose from the creodonts, that of the fissipeds which has radiated into all the modern types of land carnivores (such as dogs, raccoons, bears, civets, cats, and hyaenas). This radiation got under way in the Eocene and the fissipeds then began to replace the creodonts. The greater part of the replacement occurred relatively rapidly from Eocene to Oligocene. Thereafter the creodonts survived only in a few specialized and relic forms and these, too, finally became extinct.

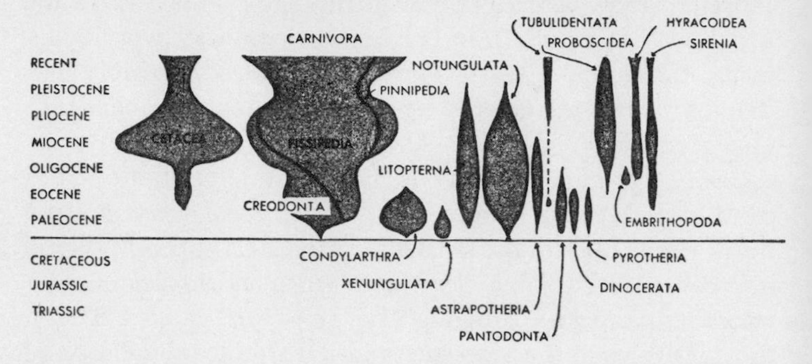

Fig. 12. Continuation of the historical record of the mammals from figs. 10–11.

Meanwhile the radiation of the fissipeds continued, especially in the Miocene, and came to include not only replacements of the creodonts but also distinctly new adaptive types of carnivores in the Miocene-Pliocene carnivore climax. Most, but not all, of these types continued in varying form to the present time. In the Miocene also appeared the pinnipeds, the seals, walruses, etc. They were a part of the Oligocene-Miocene adaptive radiation of the fissipeds, but arose by a radiating line that departed so widely from the basic fissiped type as to constitute the origin of a new structural base. As is generally the case with changes so radical, this also involved invasion of a major environmental zone, that of oceanic shores and waters, new to this group. Pinniped radiation added to the Miocene expansion of the Carnivora as a whole, and the group has continued—probably with some diminution since the Pliocene—to the present time.

Artiodactyls (even-toed ungulates) first appear in the record at the beginning of the Eocene and by the end of that period they had already increased and diversified to such an extent as to constitute a major element of mammalian faunas throughout the Old World and in North America. The Eocene forms were mostly nonruminants. Living nonrumi-

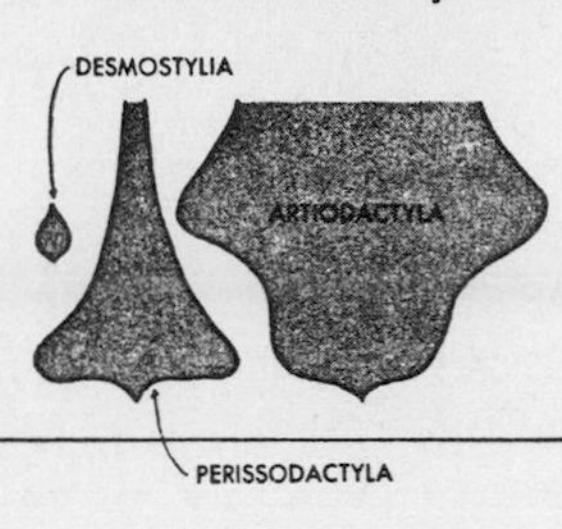

Fig. 13. Continuation and completion of the historical record of the mammals from figs. 10–12.

nant artiodactyls include pigs, peccaries, hippopotami, and camels. All of these are highly specialized surviving lines and the more varied early Tertiary nonruminants included the primitive basic type with numerous variations and also other, now extinct, lines of an adaptive radiation within this general group. Survivors of this early radiation continued to be important in faunas throughout the Tertiary, but they dwindled rather steadily and have now sharply decreased since the Pliocene.

From primitive nonruminants, one radiating type gave rise to the basic structure of the ruminant grade, members of which first appear in the record toward the end of the Eocene. This successful type in turn radiated markedly in the Oligocene and Miocene, with some replacement of the declining nonruminants, but mostly with the production of new types of artiodactyl herbivores. Recent survivors from this radiation include the little chevrotains, which closely resemble the old, basic ruminant type, the still large and far-flung deer family, the okapi and its more specialized relative the giraffe, and the unique and beautiful pronghorn or so-called "antelope" of our western plains.

Also in the Miocene and as part of the ruminant radiation arose the cattle family, Bovidae, which soon had a great expansion of its own, involving a radiation too but one of smaller scope and lesser scale than that of the ruminants as a whole. The great Pliocene expansion of the artiodactyls was due mainly to the diversification of this one family. Since the Pliocene the family has decreased somewhat in numbers of genera but it is still a dominant mammalian group and constitutes a clear majority among living hoofed herbivores. Living members include not only the cattle, for which the family is named, but also the very abundant and diverse true antelopes of the Old World, the true buffaloes, our so-called buffalo (properly bison), our mountain "goat" (which is really more of an antelope and related to the European chamois), the true goats, the sheep, and the musk ox.

Throughout their history the artiodactyls, most successful of hoofed herbivores, have been in contact with other orders of hoofed herbivores and all the members of this important, broad mammalian adaptive type have interacted widely, in part by competition, in part by inhibition of incipient radiation of one group or another. Most important of these hoofed mammals, other than artiodactyls, were the perissodactyls (odd-toed ungulates), now sparsely represented by the rhinoceroses, the tapirs, and the horses (which are practically extinct in the wild state) and their close allies the donkeys, onagers, and zebras. In the Eocene the perissodactyls were even more abundant than the artiodactyls and were the dominant hoofed herbivores. This exuberance suggests the beginning of an adaptive radiation of major dimensions, but the radiation that actually ensued was really of limited extent and followed only a few rather sharply defined lines. After the Eocene, the perissodactyls declined rather steadily throughout the Tertiary and into the Recent. What evidently happened was that the incipient perissodactyl radiation was inhibited and restricted to few possibilities by the remarkably successful artiodactyl radiation, which somehow kept a jump ahead of that of the perissodactyls. As the perissodactyls declined there also seems to have been some definite ecological replacement of them by artiodactyls.

In such histories there is another important complicating factor which also requires brief exemplification. It has some influence in the history of every group of animals, but for present purposes the general nature of this influence can be well suggested among the hoofed herbivores. This factor is that of geography, of the distribution of the various animal groups on the face of the earth and the evolving environmental features on which this distribution largely depends.

The basic type of hoofed herbivores in general appeared at the very beginning of the Tertiary, in the early Paleocene,[7] in a then abundant group of animals called condylarths (see fig. 14). These animals spread all over the continental lands of the earth except, probably, Australia, where there are not and probably never have been any wild hoofed mammals and which does not enter into this part of history.[8]

7. A group appearing at the very beginning of the Paleocene was almost certainly becoming distinct in the latest Cretaceous, and it has been claimed that a recently discovered late Cretaceous mammal was indeed already a condylarth. It is, however, so extremely primitive as to be lacking in characteristics really definitive of the Condylarthra, and its affinities are uncertain.

8. No Paleocene or Eocene land fossils are known from Australia; so far as the record goes anything or nothing may have occurred on the continent at that time. Yet if hoofed herbivores of any sort had ever reached there, there is no apparent reason why some of them should not

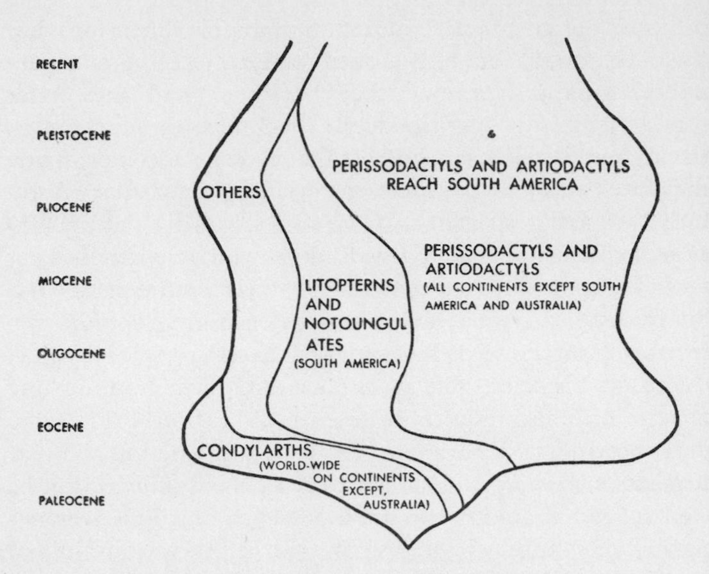

Fig. 14. The historical record of hoofed herbivores, shown as the sum of the records of various particular orders of animals of this adaptive kind.

At some unknown place and at a time probably toward the end of the Paleocene, condylarth radiation produced the new structural grades basic to the perissodactyls and artiodactyls. These two groups quickly spread in the Eocene and soon accomplished the now familiar processes of replacing and eclipsing most of the other products of condylarth radiation and extending their own radiations beyond this earlier scope. Condylarths soon (mostly by the end of the Eocene, the last rare survivor in the Miocene) became extinct as such, that is, their less divergent radiating lines all disappeared. A few

---

have survived and it seems fair enough to guess that they were never there. Other indirect considerations support this guess and perhaps justify giving it the dignity of a theory.

other sharply distinctive orders did arise from this general base, but these (simply included as "others" in fig. 14) are not essential to the present story.

The perissodactyls and artiodactyls failed, however, to reach South America. The migration route used by the condylarths to reach South America in the Paleocene sank beneath the sea and left the continent isolated, with no way of entrance for the spreading Eocene land mammals of the rest of the world. Hoofed herbivore radiation continued in South America, but separately from that of the other continents. Instead of perissodactyls and artiodactyls, it produced animals rather similar, many of them, in habits and general appearance, but quite distinct in ancestry and in details throughout their anatomy. These typically South American animals are, for the most part, the members of two orders called Litopterna and Notoungulata. In South America the litopterns and notoungulates did much the same as did perissodactyls and artiodactyls elsewhere in the world: they replaced the less radically evolving lines in the condylarth radiation so that condylarths, as such, became extinct, and they evolved and radiated farther into a large and varied fauna of hoofed herbivores.

Toward the end of the Pliocene the Isthmus of Panama arose from the sea. Some North American perissodactyls (horses and tapirs) and some artiodactyls (peccaries, camels, and deer) migrated to South America by this route, thrived there, and during the Pleistocene replaced the litopterna and notoungulates, which became extinct.

This is far from being the whole story, which, as usual, is too complex for such facile generalizations and is not fully known anyway; but as far as it goes it is a reasonable inference as to an essential part of what happened in the history of the hoofed ungulates. It illustrates phenomena and evolutionary factors such as may also be seen in the history of most, indeed of all, groups of animals.

VII. THE HISTORY OF THE PRIMATES

Enquiry into the meaning of evolution to us humans suggests special attention to the evolution of the mammalian order to which we belong, that of the Primates.

Because intelligence is our own most distinctive feature, we may incline to ascribe superior intelligence to the basic primate plan, or to the basic plan of the mammals in general, but this point requires some careful consideration. There is no question at all that most mammals of today are more intelligent than most reptiles of today. I am not going to try to define intelligence or to argue with those who deny thought or consciousness to any animal except man. It seems both common and scientific sense to admit that ability to learn, modification of action according to the situation, and other observable elements of behavior in animals reflect their degrees of intelligence and permit us, if only roughly, to compare these degrees. In spite of all difficulties and all the qualifications with which the expert (quite properly) hedges his conclusions, it also seems sensible to conclude that by and large an animal is likely to be more intelligent if it has a larger brain at a given body size and especially if its brain shows greater development of those areas and structures best developed in our own brains. After all, we *know* we are intelligent, even though we wish we were more so.

Most modern mammals exhibit more modifiable behavior and most of them have more manlike brains than do reptiles, and mammals are therefore now more intelligent than reptiles, on an average, at least. But what we know of early mammal brains, which is not much but is enough to be indicative, suggests that this was not a property involved in the origin of mammals. The little eohippus lived about 125,000,000 years after the mammals arose and it belonged to one of the specialized, "modern" orders of mammals and to the same family as its descendant the horse, which is a very brainy animal as nonprimates go. Yet the brain of eohippus was on about the same structural grade as that of a modern reptile,[1] and eohippus cannot have been much, if any, brighter than its reptilian forebears in the Triassic. Even some Recent mammals have brains and behavior patterns that imply little better than reptilian intelligence: the opossum or the armadillos, for instance. A typically mammalian level of intelligence, then, is not an original or early characteristic of the Mammalia, but has developed independently and rather differently in each of the different groups of mammals that show it. We do not know how many such independent increases of intelligence there have been, but apparently a large number. The case of eohippus shows that the development must have been independent in each *family* of perissodactyls and not only in the Perissodactyla as a whole. Of course it does not follow that each such development of intelligence among the various mammals has gone in the same direction or has reached a comparable degree. In fact it is perfectly clear that such is not the case. The generalization remains that the Mammalia, as a rule with its exceptions, have tended to become more intelligent in the later part of the history, during the last 65,000,000 years or so.

Primates appear in the Paleocene,[2] but unfortunately as

1. See T. Edinger's monograph, *Evolution of the Horse Brain* (Geol. Soc. Amer., Memoir 25 [1948]), which is cautious in conclusions and highly technical in expression but which has important implications for the study of the evolution of intelligence.

2. Occurrence of a late Cretaceous primate has recently been claimed, but as in the case of the supposed Cretaceous condylarth (see footnote 7

yet no fossil primate brain has been found in the Paleocene.[3] The oldest primate brains now known are from some time along in the Eocene, when considerable evolution beyond the Paleocene forebear is likely. The Eocene specimens show brains definitely better than reptilian but considerably below most Recent mammals even outside the Primates. Using another approach, it is most improbable that any living primate is less intelligent than the first primates, and some of the living forms are very stupid, indeed. Some of them have decidedly simpler brains, and are presumably less intelligent, than the known Eocene forms. The living tree shrews, which are either very unprogressive primates or equally unprogressive descendants of the immediate ancestors of the primates (a distinction in classification that really makes no difference), have brains that are among the simplest known in living mammals and that are not enough above the reptilian level to matter much.

The reasonable conclusion is that the first primates were not characterized by advanced intelligence but on the contrary were in this respect about as low as any of their fellow mammals and only a little better off than the Triassic reptiles. Intelligence developed among primates in the course of their later evolution. It developed unequally in different lines, in some hardly at all, in others to the highest degree ever reached by any being. The primates, as they turned out and not as

---

of the preceding chapter), its known characters are so primitive, so lacking in reliably diagnostic primate characters, that its classification as a primate is at present highly dubious.

3. Strictly speaking there is no such thing as a fossil brain, because the brain tissue itself never fossilizes. What is preserved is a cast of the inside of the skull. From this an expert can reconstruct a reasonable approximation of the brain. The distinction involved and the methods used are explained in the monograph by Edinger cited in footnote 1 of this chapter.

they arose, are distinguished by *including* the brainiest animals, not by *being,* as a whole, the brainiest. Typical developments of primate intelligence were independent from other developments of mammalian intelligence and were different. If we could measure equal quantities of intelligence in, say, a dog and a monkey (which we can't, but we can imagine doing so), the qualities still would be quite unlike.

The primate variety of intelligence, although of course it differs in quality also among different primates, is apparently connected closely with main dependence on the eyes for information about the outside world and with use of the hands (or forefeet) to manipulate objects. These are undoubtedly basic characteristics of the first primates, even though the intelligence now related to them is not. Both characteristics also seem to have been related to life in trees: you need good vision and grasping hands to climb. The basic primate also had rather simple teeth fit to cope with almost any sort of food that did not make unduly heavy or special demands on the dentition. In fact at this level the primates were barely distinguishable from the basic placental mammal or nominal insectivore. No peculiar or evolutionarily potent structural or physiological adaptation was involved in their origin, no apparent new grade or type of organization such as is usually seen in the origin of a group that later rises to importance. Among the primates such distinctions came later, and only to some of the branches in their radiation.

Primate classification has been the diversion of so many students unfamiliar with the classification of other animals that it is, frankly, a mess. It involves matters of opinion on human origins and, humans being what they are, such opinions are endlessly varied and not always distinguished by competence or logic. Fortunately, however, the technical details need not concern us here and many students of the group will quarrel only halfheartedly with the following simplification of the problem:

Broadly speaking, four general levels and types of structure can be distinguished among the primates, although they are of unequal fundamental variety or taxonomic value. (See fig. 15.) First is a group including the oldest known primates and a great number of anciently divergent lines from them which have in common little more than the fact that they do

Fig. 15. Representatives of the main groups of primates. The lemur represents the prosimians; the capuchin monkey, the South American monkeys or ceboids; the macaque, the Old World monkeys or cercopithecoids; the chimpanzee, an ape, and the man both represent the broadly manlike or hominoid group. Each group includes a variety of different forms, only one of which is shown as an example.

not lead directly into any of the other three groups and that in general they have been unprogressive in brain evolution and some other respects. These may be called the prosimians and they include, among now living animals, the lemurs, indris, aye-ayes, lorises, pottos, galagos, tarsiers, and, with some reservations, the tree shrews, which may be primates or may only be the next thing to it.

A second group is that of the South American monkeys or ceboids, after *Cebus,* the technical name of what used to be the usual organ-grinder's monkey. They are neatly packaged by the fact that they live in South America, are the only monkeys that live there, and have never lived anywhere else—except in recent times in the faunal extension of South America through Central America into tropical Mexico. More anatomically but more frivolously, they include the monkeys with tails by which many of them hang. It also happens that they have broad, flat nose cartilages. Among them are the marmosets, howling monkeys, capuchins, spider monkeys, and others.

The third group is that of the Old World monkeys or cercopithecoids—*Ceropithecus* being the common African monkey, not baboon or ape. They share with the South American monkeys, but by independent development, more evolved brains than those of the prosimians and a somewhat more human look due mainly to their usually more flattened or shortened faces. They usually have tails but cannot hang by them and their nose cartilages happen to be relatively narrow. The macaques and rhesus monkeys of medical experimentation belong here, and so do the baboons, mandrills, guenons, langurs, and other monkeys throughout Africa and southern Asia and barely into Europe at Gibraltar.

The fourth group, automatically highest in our view, is that of the apes and men or hominoids (named by *Homo* after himself). They have no tails, are relatively large in size, and exhibit in various degrees the highest development of intelligence yet attained. The living forms are the gibbons (several species), orangutan, chimpanzees (two species), gorilla, and man.

The known fossil records of these groups are summarized in figure 16, separately by continents (and Madagascar) and then combined for the world as a whole. Australia is left out because the only native primate that it has, or is likely ever to have had, is man. The fossil record is none too good, inferior to that for most orders of mammals, and would hardly merit special attention here if it were not the record of our own order so that we are inclined to make the most of it. The total record at once shows that much is missing, for it has some of the groups pinching out altogether for different lengths of time when previous and later records show that the animals must, nevertheless, have been fairly common.

To make the most of this faulty record, an attempt may be made to reconstruct from it the facts that it portrays with obvious infidelity. There are some clues to such reconstruction that remove it from the realm of pure guesswork. First, the Recent record is a given fact. It is not likely that anyone will ever discover a now unknown genus of living primates, and another genus or two would hardly show on this scale, anyhow. Now for the prosimians we have a large number of Pleistocene and Recent genera, mostly on the island of Madagascar, and an even larger number of Paleocene and Eocene genera, but little in between. The Paleocene and Eocene records are probably fairly good for limited parts of North America and of Europe. Appropriate fossil deposits of this age are almost unknown in Asia and Africa but indirect evidence strongly suggests that prosimians were there at that time (but not in South America or Australia) in numbers comparable to those of Europe and North America. It seems conservative to estimate that there were at least twice as many Eocene genera as those we know; and the discrepancy for the Paleocene, with a manifestly poorer record even in

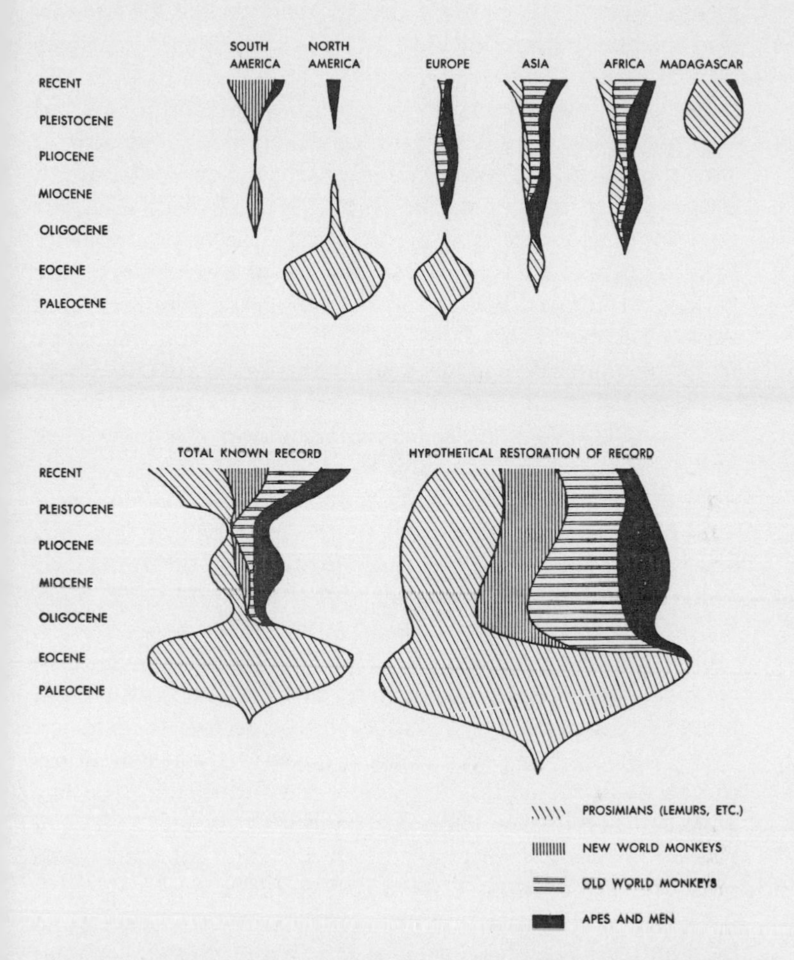

Fig. 16. The historical record of the primates. The record is constructed like those of figs. 10–13. The upper part shows the known record separately for geographic regions and the lower figure shows the total known record for the world and an attempt to restore the real variety represented by the partial record, as discussed in the text.

regions where it is known, must be even greater. The reduction and final disappearance of prosimians from the later Tertiary records of Europe and North America is probably in accordance with fact and apparently represents a real and strong reduction in their area of distribution and also in their numbers. It is hypothetical how far this reduction went, but probably little or not at all below the Recent level, for the prosimians seem now to be a waning group in spite of their fairly rich survival in the asylum of Madagascar. The paucity of known records for the Oligocene, Miocene, and Pliocene is explicable if, as is likely, prosimians were then largely confined to Madagascar and to tropical areas in Africa and Asia (including its islands), where pre-Pleistocene fossils are poorly known.[4] So the prosimian part in primate history as reconstructed in the graph to the right in figure 13 is a reasonable rather than a wild guess.

As for the ceboids, there is every reason to believe that they have always been confined to tropical South (and more recently also Central) America, except for strays into more temperate parts of the same continent. The whole pre-Pleistocene record for this region is poor, so the paucity of primate records is meaningless. Discoveries of a few fossil ceboids in both temperate and tropical South America do give the useful fact that they were there, and fairly typical in structure, toward the end of the Oligocene and in the Miocene. Their most probable ancestors are certain North American Eocene prosimians, so origin around the end of the Eocene or in the earliest Oligocene is reasonable. All native South American animals with better records reached a maximum by middle or late Tertiary and it is likely that the ceboids

4. Some rather rich faunas are now known from tropical Africa, and they do contain prosimians, related to the living lorises and galagos. Although most prosimians died out in North America by the end of the Eocene, extremely rare, aberrant stragglers did continue into the Oligocene and Miocene.

did too, but that their decline has been slight, as they are still varied and successful animals.

The record of the Old World monkeys suggests origin (probably from Old World prosimians) in the late Eocene. The poorness of the Miocene[5] and Pliocene record is the main deficiency of knowledge here, and it is a reasonable inference that this group expanded along with associated Old World faunas of more or less modern type at that time. There has probably been some Recent reduction, but the record suggests that this has been slight, as few radically distinctive extinct Pliocene and Pleistocene forms have turned up.

The hominoid record is surprisingly good, with fewer complete breaks than in the cases of the other groups. Still there is a gap in the African Pliocene record, and earlier and later fossils from that continent show beyond much doubt that it was a major center for this group in the Pliocene, also. The known Miocene and Pliocene hominoids are not likely to represent as many as half those actually existing at those times, even though we already know a greater variety of hominoids in those epochs than exists today.

This reconstruction of the record, and the data on which it is based, seem to give a fairly clear picture of the outlines of primate evolution as a whole. (See fig. 16.)

The first primates were prosimians, which quickly spread over most of the world. They underwent a remarkable adaptive radiation in the Paleocene and Eocene, when they swarmed in Europe and North America, at least, and doubtless far more widely. Among the multitude of still closely related but rapidly diverging lines of that time, the majority became extinct by late Eocene or early Oligocene, when the whole group suffered a reduction in geographic extent perhaps connected with slow climatic changes. (This is hypo-

5. The Miocene African faunas mentioned in the preceding note include monkeys, but they have not been adequately described at the present writing. A few Old World Pliocene monkeys have been described.

thetical, but there is some evidence that zoning of climates gradually became more pronounced after the Eocene and that what are now the temperate regions slowly became cooler, on an average.) There was perhaps some replacement by monkeys in the Old World, but on the whole this reduction seems to have been rather by displacement than by replacement. Indeed among the primates as a whole, displacement and expansion into new ecological positions seem to account for contractions and expansions with relatively little replacement of one group by another.

A few of the diverging early prosimian lines did survive in the Old World. Although they remained primitive on the whole, unprogressive in brain structure and some other respects, these lines had some peculiar specializations in one part or another of their anatomy. In many of them the lower incisor teeth became long and pressed closely together, forming a comblike structure. In several, notably the tarsier, there was a peculiar lengthening of some of the ankle bones and development of an extra joint in the hind foot. The tarsier remained arboreal but became nocturnal and has the additional peculiarity that the eyes are relatively enormous, point straight forward, and seem to occupy almost the whole face—a development adaptive for an active animal that moves about in the trees at night. In fact most of the now surviving prosimians are oddly specialized in one way or another.

As a peculiar side issue of primate history, some time during the Tertiary a group of prosimians managed to reach the island of Madagascar. Here they found an asylum where they had few enemies and no competitors and they underwent a new adaptive radiation confined to that large island. This radiation probably culminated in the Pleistocene, although we have no earlier primate fossils from Madagascar and earlier forms could perhaps have been even more varied. The radiation included some very strange developments, such as a lemur, the aye-aye, with rodentlike gnawing teeth[6] and another as large as a man. (As a rule, the prosimians, fossil and recent, are small animals, generally ranging from the size of a rat to that of a fox terrier.) This Madagascar radiation largely accounts for the odd fact that prosimians are still more fundamentally varied than any other group of primates.

Another line of early prosimians similarly invaded the larger island (as it was then) formed by the continent of South America and similarly underwent a separate radiation there. The result was different, however, in that these animals did not simply radiate on the basis of the primitive prosimian level but advanced to what deserves to be recognized as a distinct grade of primate structure. A feature of this change was modest but definite brain enlargement, with correlated remodeling of the skull so that the snout, long in many prosimians, became short and the eyes, primitively lateral in prosimians, came to occupy much the same position that they do in man.[7]

6. Rodentlike enlargement of the front teeth was also one feature of the Paleocene-Eocene prosimian radiation and then developed in several different lines. Earlier students believed that the aye-aye is a direct survivor of one of these lines, and this idea is still occasionally given in discussions of primate evolution. (It often takes a long time for new theories to replace the old in textbooks!) Recent studies demonstrate that it is much more probable that the aye-aye is a new development in the Madagascar radiation which happened to parallel some of the products of the older radiation in this respect.

7. The prosimian tarsier also has a short face and anteriorly directed eyes, as noted above, and so to less degree did some Eocene prosimians. Although it is sometimes hailed as evidence of special relationship to higher forms, this development is almost certainly quite different and independently evolved. Enormous relative enlargement of eye size in these prosimians, not paralleled in higher primates, simply overshadowed the snout and, so to speak, crowded it out of the face down under and between the eyes.

In the Old World two other lines of the early Tertiary prosimian radiation also progressed to the point of representing new grades of primate structure, initiating in each case new radiations extending into new ways of life: the Old World monkeys or cercopithecoids and the hominoids. These appear in the fossil record at the same time, and there is every reason to believe that they did, in fact, originate at about the same time. They may have arisen from two different lines of prosimians, and the ceboids from a third. Thus the four main primate groups do not (although many early and a few recent students have thought they do) represent four successive steps, each leading to the other, like, on a larger scale, Osteichthyes to Amphibia, Amphibia to Reptilia, and Reptilia to Mammalia. Prosimians are on the whole and in some respects more primitive than other primates and they did clearly arise first. Ceboids are in a few rather minor anatomical details (such as the number of premolars) more primitive than cercopithecoids, and cercopithecoids are in some ways (such as retention of the tail or the usually lesser intelligence in Recent forms) more primitive than hominoids, but these three groups arose at about the same time and not one from another.

The prosimians apparently gave rise separately to each of the three other groups, although the ancestral hominoids and cercopithecoids were more closely related to each other than to the early ceboids. In their basic forms, the three groups can be arranged in a sequence of "primitiveness" or "progression" from "lower" to "higher" only by a more or less arbitrary choice of characteristics, and sometimes not even by this device unless we eliminate ancestral types. The recent hominoids are certainly the highest forms if we define "highest" as "brainiest," but there is no evidence that the earliest hominoids, or the forms in the transitional prosimian-hominoid line, were notably more intelligent than the contemporaneous prosimian-ceboids or prosimian-cercopithecoids. They may have been, but this is not particularly probable. If we use other criteria for "higher" and "lower," the sequence may be radically different. The prosimian aye-aye, for example, is more distant from the ancestral condition, and hence may be said to be "higher," than is man in some respects.

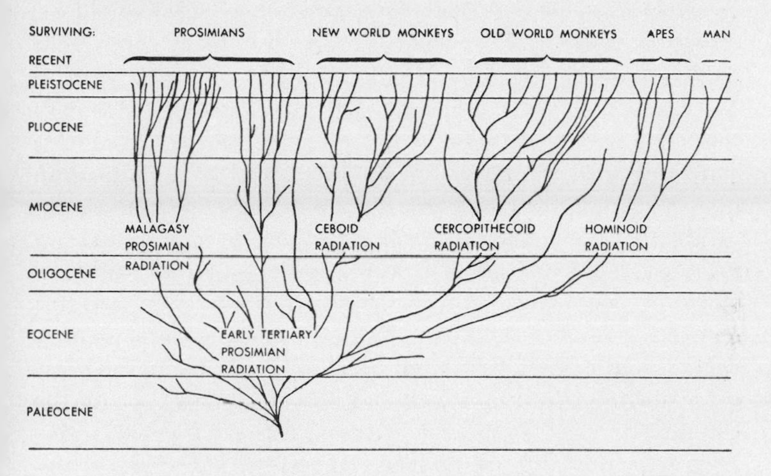

Fig. 17. A representation of primate history and the origin of man as a series of adaptive radiations in space and time. (The selected lines of descent suggest a theoretical over-all pattern, but most lines are omitted and those shown are simplified and generalized.)

A diagrammatic representation, extremely simplified and yet, I believe, true in essential outlines, might show primate history in terms of lineages and radiations more or less as in figure 17. There is a highly diverse basic radiation of the first primates with a few surviving lines that have not themselves re-radiated significantly and that are, for purposes of classification (which always must have some arbitrary elements), united with the early forms as prosimians. Another line re-radiated in Madagascar on a scale of minor importance and is still called prosimian. A line radiated in South America and produced the ceboids. Two others radiated in the Old World and produced cercopithecoids and hominoids.

In some respects the cercopithecoid radiation in the Old World paralleled that of the ceboids in South America, in a way rather distantly analogous to artiodactyl-perissodactyl radiation outside of South America and litoptern-notoungulate radiation in that continent. Some similar adaptive types occur in each radiation. Ceboids include some very small arboreal types not closely paralleled among cercopithecoids in the Old World, where there were still some prosimians of this general adaptive type. The cercopithecoids also produced a striking new type of large, ground-living primates, the baboons, that did not appear among the ceboids.

Among the hominoids,[8] the group eventuating in apes and men, at least one divergent branch seems already to have arisen in the early Oligocene in *Apidium* from Egypt, a relative and possible ancestor of *Oreopithecus,* from the Pliocene of Italy. That lineage paralleled the one leading to man in some respects and has therefore been considered more closely related to the latter, but it now seems fairly certain that the resemblance is in fact due to parallelism and that the relationship is distant. The *Oreopithecus* lineage is extinct. The line leading to the several living gibbons

8. The following are a few of the many studies of recent and fossil hominoids and the biological origin of man:

W. E. LeGros Clark, *History of the Primates* (London, British Museum [Natural] History, 1956, and *The Antecedents of Man* (Edinburgh, The University, 1959; also in paper as a "Torchbook," New York, Harper and Row, 1963). J. Buettner-Janusch, editor, "The Relatives of Man" (*Annals New York Acad. Sci.*, 102 [1962], 181–514). S. L. Washburn, editor, *Classification and Human Evolution* (Chicago, Aldine, 1963). W. Howells, *Mankind in the Making* (Garden City, Doubleday, 1959), and as editor, *Ideas on Human Evolution* (Cambridge, Harvard Univ. Press, 1962).

of southeastern Asia may also have split off in the Oligocene and had certainly done so early in the Miocene, as attested by fossils, some nearly complete, from Europe and Africa. These are the most specialized arboreal apes, with enormously long arms; the fossils show that this specialization was acquired rather rapidly and late in the history of the group.

In the Miocene and into the early Pliocene there was a complex adaptive radiation of the hominoids, represented for the most part by tantalizingly fragmentary fossils from Africa, Europe, and Asia, including *Proconsul,* the best-known genus at present, and *Dryopithecus,* the commonest and most widespread: members of this not very well understood complex of Mio-Pliocene apes are commonly referred to as dryopithecines. The connections are not clear, the fossils being comparatively few and fragmentary, but the lines leading to the orang, the gorilla, chimpanzees, and man apparently all diverged in this radiation, while still other lines became extinct. There is evidence, probable but not quite conclusive, that a line leading to man was distinct from that leading to any recent ape before the end of the Miocene. Evidence not only paleontological but also or more particularly biochemical, cytological (concerning fine structure of cells and nuclei), physiological, and anatomical make it now fairly certain that the Miocene ancestry of man was especially related to that of gorilla and chimpanzees, which are still more closely related among themselves.

The australopithecines, best known from East and South Africa but perhaps occurring throughout the warmer parts of the Old World, are first known by specimens so far discovered from the early Pleistocene, upwards of 2,000,000 years ago. Although little or not at all advanced in brain size, at least, beyond the apes, they belong to the human family, not only hominoids (a term including apes) but hominids (a term excluding apes and all but our closest relatives). For a long time in the Pleistocene there were at least two and possibly three or more different specific lines of descent within this family, one ancestral to ourselves and others becoming extinct. By some time in the middle Pleistocene, perhaps 500,000 years ago or perhaps less, the hominid family was represented by only one line of descent, which eventually became the single species, *Homo sapiens,* that includes all now living hominids.[9]

A phenomenon just noted on a small scale in the hominids also occurs on a larger scale in earlier primate radiations and is, indeed, an important general factor in evolution. That is the process of weeding out, by which among multiple related lineages many become extinct, leaving only those most efficiently adapted to some particular way of life as survivors. This is a special type of replacement, occurring within a given group especially during and following a phase of increase and radiation. Such a phase is usually initiated when a population either enters a new geographical environment or, by structural and physiological changes, is enabled to expand in a new ecological zone. The first reaction is commonly the production of a great multiplicity of highly varied, markedly variable, but still rather similar groups, subspecies, species, genera, and finally perhaps incipient families on the taxonomic scale. The process of weeding out then begins. Among closely similar variant groups one or two begin to attain dominance over the others and most of the variants are

9. This has been the briefest possible account of what now (1966) seem to me the most probable, best-evidenced, broad outlines of human origins. This account follows a strong consensus, but of course there are dissenters, and some details, comparatively unimportant, remain quite uncertain and disputed. The purposes of the present book prevent going into such details or even exploring such important questions as those of race. For the general biological problems of species and race E. Mayr's *Animal Species and Evolution* (Cambridge, Harvard Univ. Press, 1963) is *the* authority, and for their application to the human species Th. Dobzhansky's *Mankind Evolving* (New Haven, Yale Univ. Press, 1962) is unequaled.

eliminated. Within single populations, too, the type tends to become more constant and variability usually tends to decrease. There seems to be a premium on divergence of the various populations from each other but on development of a well-fixed and particular type within each. In time the multiplicity of the ancestral group is reduced to a smaller number of more sharply distinct adaptive types, each of which tends to specialize in some particular way of life.

One of the most fully documented examples of this process is that of the extinct notoungulates of South America, the radiation of which was of major scope and yet took place on a single, isolated continent where it can be well followed.[10] In the Eocene these animals are found in the first bloom of their expanding phase, a phase commonly called "explosive" by students of evolution, although considerable imagination must be used to conceive of an explosion that makes no noise and goes on for several million years. Within the Eocene in a single region (central Patagonia), still imperfectly explored, seven families and (at least) twenty-seven valid genera of notoungulates are known. Although distinctive in detail, these are all similar and intergrade to such an extent that their identification requires closest study. Moreover, within single species the amount of structural variation is extraordinarily great. By the early Miocene, with loss of some families and differentiation of some, the net number is five and these are represented by only seven rather stereotyped, widely distinct genera. These

10. See G. G. Simpson, "History of the Fauna of Latin America," in *Science in Progress,* Seventh Series (New Haven, Yale Univ. Press, 1951), also G. G. Simpson, *The Geography of Evolution* (Philadelphia, Chilton, 1965). The technical data for the basic radiation and beginning of the weeding out process are given in G. G. Simpson, "The Beginning of the Age of Mammals in South America" (Part 1, *Bull. Amer. Mus. Nat. Hist., 91* [1948], 1–232; Part 2, in press in same bulletin for probable publication in 1967).

early Miocene lines continue, with some progressive change and secondary diversification but no essential modification of adaptive type, into the later Tertiary where one by one they drop out, with only two families and four genera reaching the Pleistocene. In the latest Tertiary and Pleistocene the notoungulates were definitely waning and being replaced by forms from other orders, but before that time the reduction in variety represents mostly the segregation and specialization of a few types each of which pre-empted a way of life incipient for a greater variety of the Eocene forebears.

In such a case, the reduction in numbers of genera or other units does not necessarily mean a reduction in numbers of individual animals. It may even mean an increase in numbers of individuals, with replacement of many small competing population groups by a few successful and well-adjusted groups each of vast abundance and not competing significantly with each other. Thus we have the apparent paradox in evolution that the success of a group of animals may result in (or, at least, accompany) a marked decrease in their variety. This does not always happen and it depends on definition of "success," but it seems sometimes to happen if success is measured by abundance of individuals.

In the less well-documented history of the primates, it seems to be some such process as this that is operating when each radiation is followed by a reduction in number of lines deriving from it and when these lines tend to diverge, to specialize independently each in its characteristic and often increasingly stereotyped way. Man is the end, so far as the end has yet been reached, of a line from the Miocene hominoid radiation. Other lines have diverged and specialized as the gibbons, orangutan, chimpanzees, and gorilla, following paths involving little competition with each other or with early man. Other ancient lines, near one or the other of these at various phases of their history, have dropped out as the successful (that is, the surviving) types became established.

The divergent australopithecines, and perhaps others at different times in the history, are probable among those whose way of life was pre-empted by the line leading to man. The most successful type, to this point, belongs to a single genus, *Homo,* and to a single species, *Homo sapiens,* but he is certainly far more abundant in individuals than any hominoid has ever been before. Since the Miocene, the variety has much decreased but the abundance is at its highest point and still increasing—all too rapidly!

VIII. RATES OF EVOLUTION

An important aspect of the record of evolution, implicit in much that has already been said, is the rate at which it has occurred. On examination, this subject turns out to be unexpectedly complicated. Half the problem in learning about anything is asking the right questions. To ask "How fast has evolution occurred?" seems like a simple and proper approach to the present subject, but this is not the right question. In the first place, it must already be evident that there is no such thing as *the* rate of evolution. The record has demonstrated that evolution is not some over-all cosmic influence that has been changing all living things in a regular way throughout the periods of the earth's history. Some groups have changed rapidly while others were remaining practically unchanged. The same group is commonly seen to have changed rapidly at some times in its history and slowly or not at all in others. Within a given group some parts may change while others are static. So the question "How fast has evolution occurred?" is meaningless unless we add, "The evolution of what group of organisms, of which of their structures, and at what time in their history?" These variations in rate are in fact important in themselves and should teach us something about the meaning of evolution.

In the second place, evolutionary change has several different aspects which are not identical or even comparable in the rates at which they occur. Little eohippus evolved into the horse. This sort of structural change from an ancestral type to a more or less markedly different descendant type is, of course, evolution and it has a rate, so much change per geological epoch or per million years or per thousand generations. As a matter of fact, all sorts of new difficulties appear when we attempt to measure or even to define this rate, but it is one definite kind of answer to the question of how fast evolution occurs. Then again, it is certainly evolutionary change when one species splits into two, or one family into twenty-three (as we saw the ammonites do from Triassic to Jurassic). This sort of evolutionary process accounts in considerable part for the rise (and its cessation for the fall) of different groups of animals as we have followed them through geologic time. Its speed, in the various cases at various times, is a rate of evolution, and obviously a very important one. It is, however, a different sort of rate from the rate of structural change and it is almost as hard to compare the two, even though they are both related parts of the total process of evolution, as it is to multiply apples by oranges.

There are other sorts of evolutionary rates, but for present purposes at least the two most important are rates of structural change and rates of diversification. The technical difficulties in measuring either one with any accuracy are so great that there are not many good measurements yet available from the record, but a few suffice to give some insight into their bearing on the meaning of evolution.[1]

1. I considered this subject in some detail in "Evolutionary Rates in Animals," in the symposium *Genetics, Paleontology, and Evolution* (Princeton, Princeton Univ. Press, 1949), also in *The Major Features of Evolution* (New York, Columbia Univ. Press, 1953), and elsewhere. Among numerous shorter technical studies of evolutionary rates are several by B. Kurtén, for example, "A Differentiation Index, and a New Measure of Evolutionary Rates" (*Evolution, 12* [1958], 146–157), and "Rates of Evolution in Fossil Mammals" (*Cold Spring Harbor Symp. Quant. Biol., 24* [1959], 205–215). Most recent works on evolutionary principles include brief reviews, at least, of this subject, notably B. Rensch, *Evolution above the Species Level* (New York, Columbia Univ. Press, 1960), for higher categories, and E. Mayr, *Animal Species and Evolution* (Cambridge, Harvard Univ. Press, 1963), for lower categories.

The amount of total structural change from eohippus to the recent horse, for instance, can be estimated by the number of genera into which competent students divide the direct lineage. This number is, of course, subjective because the series is continuous and there is no natural division of it into genera or other units.[2] There are eight genera so recognized, omitting the last, *Equus,* because it is still going and we do not know how long it will take to evolve into another genus, if it does. The change involved in progression through these eight genera took about 50,000,000 years, so we have the approximate figure for average rate of evolution in this line of horses of about one genus per 6,250,000 years. Figures for other hoofed mammals during the Tertiary seem to have about the same order of magnitude and suggest that this is a usual sort of rate for a land animal during that period.

It is nevertheless clear that some have evolved at quite different rates. The opossum lineage, for instance, is not completely known, but there are Cretaceous opossums strikingly like the living forms. In 70,000,000 years or so the line leading to the recent opossum certainly changed far less than did the line leading to the recent horse in 50,000,000 years. The rate of evolution must have been very much lower in the former than in the latter. Some figures based on Permian and Triassic ammonites, those extinct relatives of the chambered nautilus of present seas, give rates on the order of one genus per 20,000,000 years. With due allowance for the fact

2. Measurement of total change in more directly quantitative and perhaps more objective terms would have to take into account hundreds or even thousands of characteristics of different kinds, measurable in different units, and changing at different rates. That imposing problem is not wholly insoluble in principle, and some progress toward its solution has been made by using complex statistical methods and electronic computers. At present, however, although hopeful, these approaches involve unsolved difficulties, and the results so far seem to be less significant and not less subjective than those following the sort of taxonomic approach discussed in the text.

that what one student calls a genus of ammonites may not be too comparable with what another student calls a genus of horses, it still seems fair enough to conclude that horses evolved considerably faster than these ammonites.

There is thus good concrete evidence for the impression that some animals have evolved much faster than others. There is less complete but still sufficient evidence for the further generalization that the vertebrates have tended to evolve (structurally) faster than the invertebrates. Of course this is only true as a broad average over the whole of the two enormous groups of organisms. Some vertebrates have practically ceased to evolve over long periods of their history, and some invertebrates have evolved more rapidly than most vertebrates. It also seems probable that land vertebrates have, on the average, evolved more rapidly than aquatic vertebrates, but here the evidence is still rather shaky because not enough of it has yet been compiled to give a fair guess as to the averages.

There is a general impression that the "higher" groups, those that appear later in the record or that are closer to man (not a very objective criterion of "higher"), have evolved more rapidly than the lower. Now if one recent animal is structurally more distinct from an ancient common ancestor than another, then sometime during the evolution of the former there *must have been* more rapid evolution than in the case of the latter. If living mammals (the higher group at least in the sense of being the more recent) are less like the first reptile than are living reptiles, then since mammals arose they must, on an average, have evolved more rapidly than reptiles during that same time. This apparently is true, but it does not exclude the excellent chance that reptiles evolved as fast as mammals, or faster, before mammals arose. The question also arises as to the particular respects in which one recent group is less like a common ancestor than is another. Man's brain is less like that of the Eocene primates than is the brain of the aye-aye, a contemporary of man. Therefore sometime since the Eocene brain evolution has been faster in the lineage leading to man than in that leading to the aye-aye. But the aye-aye's teeth are less like those of Eocene primates, so sometime since the Eocene the aye-aye lineage has evolved faster, as regards the teeth, than has the human lineage.

This fact that different organs can evolve at different rates either in the same group of animals or in different groups can also be illustrated and proved by examples from the well-known record of the horse family. All at the same time, in the Pliocene, there are found different groups of horses, one (*Hypohippus*) with both feet and teeth little changed from the Miocene common ancestor, one (*Hipparion*) with feet little but teeth much changed, and one (*Pliohippus*) with both feet and teeth much changed. In the first, feet and teeth both evolved slowly in the later Miocene, in the second feet evolved slowly and teeth rapidly, and in the third both evolved rapidly. Such phenomena account for the observation that man is, as regards his whole body, a rather primitive (slowly evolved) mammal, in spite of the advanced (rapidly evolved) brain of which he is so proud.

It is easy to see that the rate of evolution may vary greatly at different times within a single line of descent. The opossum has scarcely changed since the late Cretaceous, but in the late Cretaceous it was one of the most advanced forms of life then existing. Its ancestors must then have evolved rapidly at some time earlier than that, and the rate must have dropped nearly to zero thereafter. There are many examples of this phenomenon. The odd horseshoe crab, for instance, which is really no crab at all but actually more nearly related to the spiders, has existed without any notable structural change for some 200,000,000 years, but it was a progressive, highly evolved animal in comparison with its contemporaries when it first appeared in the record, and in the Paleozoic its ancestors evolved rapidly.

Although they constitute a minority in the total of today's life, there are many of these strange animals that have not changed significantly for some millions of years. It seems generally to be true of them that they arose by rather rapid evolution and were advanced forms at the time when their evolution slowed down or practically stopped.

Less spectacular but more precisely measurable changes in rate can be demonstrated for particular characters. Again the excellent horse family record provides good examples. A striking change which took place in one of the several different lines of descent in this family was a marked increase in the height of the teeth. This change was going on very slowly in the Eocene, Oligocene, and early Miocene. Beginning in the middle and continuing through the late Miocene, it went on at a much more rapid rate simultaneously in several different lineages. Then in the Pliocene it slowed down again and has changed little from early Pliocene to the present time. Other characters also had limited periods when their evolution speeded up, and these periods do not correspond for different characters. For instance, the most important acceleration of horse brain evolution seems to have been around the late Eocene, when skeletal and dental evolution was not particularly fast.

So it is seen that rates of structural evolution are highly variable. Even in the staid horse family, which seems as a whole to be progressing rather steadily through the Tertiary, close examination shows the rates varying considerably. More broadly, evolution commonly seems to proceed in spurts and pauses in an apparently erratic way, even though we may be sure that some reason lies back of these changes of rate. The more examples of long-range evolutionary rates are examined, the more it appears that a constant rate over geologically long periods of time is exceptional and that marked changes in rate are the rule.

There is another way of studying evolutionary rates that

is relatively easy to carry out, although it may be even more difficult to interpret clearly. The procedure is to count how many new genera (or other units, but genera are usually the

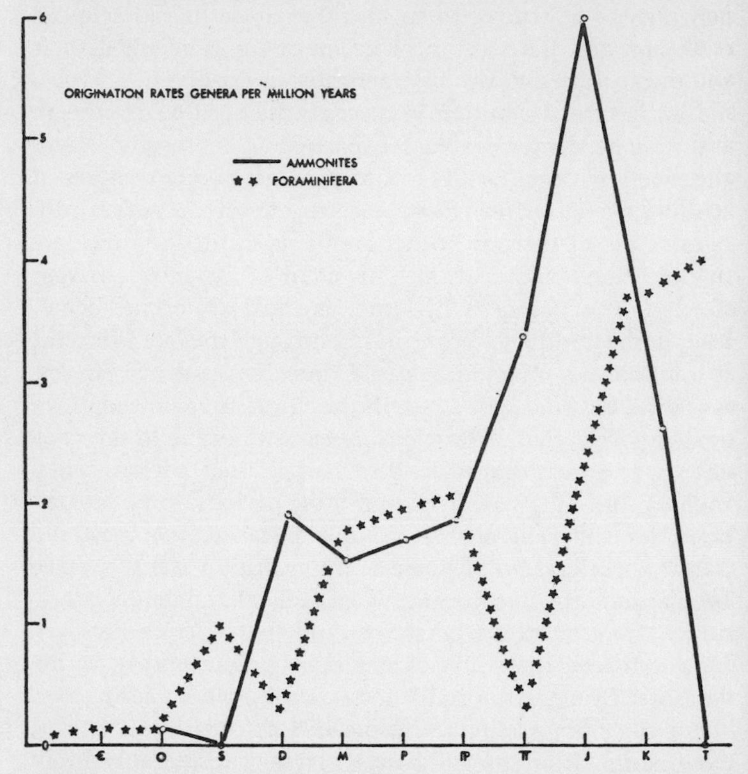

Fig. 18. Rates of evolution in ammonites and Foraminifera. The height of the curve in each period shows, against the scale to the left, the number of new genera known to appear per million years in that period. The time scale at the bottom has abbreviations for the successive periods from Cambrian to Tertiary; see fig. 1. M (Mississippian) and P (Pennsylvanian) are respectively the early and late Carboniferous of the world-wide scale. The horizontal scale is not proportional to years but arbitrarily gives equal space to each period.

most enlightening) appear in the record of a given group during some particular time and then to divide this by the length of that time span in millions of years. Of course the new genera of record do not necessarily represent all that really appeared then—no record is absolutely complete—and the span in millions of years is, as yet, only roughly measurable, but for reasonably well-recorded groups the resulting approximations are evidently close enough to cast considerable light on true rates. This rate gives the number of genera arising per million years, a rate of origination of genera.

The principal difficulty of interpretation lies in the fact that the rate is not precisely one either of structural change or of splitting up of lines of descent, a rate of diversification, but involves both at once in a way usually hard to untangle. For instance, if a single line of descent is present and it evolves so rapidly as to progress through five genera in 10,000,000 years, the rate of origination is one-half genus per million years which actually is a rate of structural change, with no diversification at all. On the other hand, one genus may persist through 10,000,000 years but during that time it may branch and give rise to five others, distinct in character but not particularly more advanced. Then the rate of origination is also one-half genus per million years, but this reflects diversification rather than structural change. In most groups the two sorts of evolution are occurring at the same time and it is difficult or impossible to be sure what part each contributes to the rate of origination of genera. Yet this latter rate does give a rough sort of idea of the total amount of evolutionary activity, of all kinds, going on in a group.

Some representative examples[3] of such rates are given in figures 18–23. It is unnecessary to describe or discuss each

3. The data for these figures are those that were available approximately twenty years before the present revision (1967). In all of these groups some new genera have been discoverd in the meantime or some genera split (often unnecessarily) into two or more. On the other hand

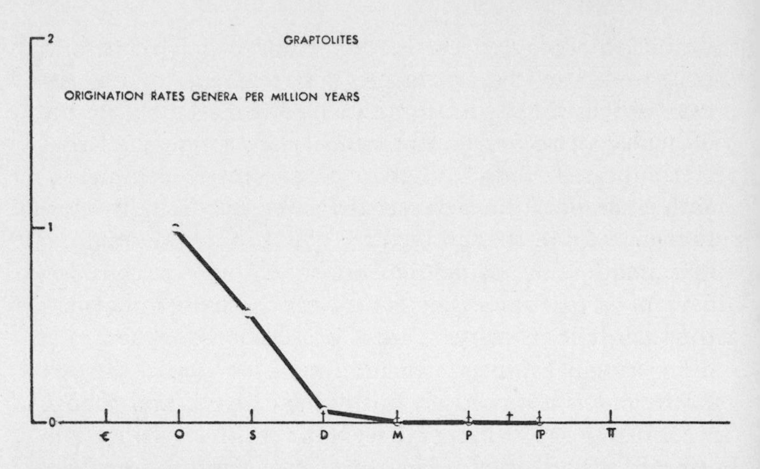

Fig. 19. Rates of evolution in graptolites. The construction and conventions of the figure are as in fig. 18. The time scale runs from Cambrian to Triassic.

example in detail, but note first of all the variety of the patterns. Some start high and drop rather steadily through the history of the group (e.g., graptolites, perissodactyls). Others start relatively low, build up to a high point, then decline rapidly (ammonites) or slowly (notoungulates) or are still high (carnivores, artiodactyls—the decline presumably lies

---

some genera then recognized have since been reduced to synonymy, and estimates of lengths of periods before the Tertiary have almost all been increased. These changes tend to some extent to cancel out. Trial for some groups with available new recent data indicates that the shape of the curve has not been significantly changed. No more can be asked of these figures and no more is claimed for them. Counts of genera were not complete in the 1940's and are not complete in the 1960's. Estimates of time in years are probably more accurate now, but are still estimates with large probable errors. Nevertheless it is sufficiently probable that the curves from data of the 1940's significantly represent the trends of change in real rates of evolution, and they are retained in the present revision.

in the future). Several show two or more quite distinct high points (ammonites, brachiopods, etc.; the three highs and two lows in the record for Foraminifera may be the result of inadequate data but are probably real).

Another important point to note is that none of the examples shows even approximately a constant level of evolutionary activity throughout the whole existence of a group of animals. There are always periods of higher and of lower activity. High points may occur near the beginning, around the middle, near the end, near both beginning and end of the span, or in other patterns, but they are always there. Highs and lows may also be either long or short in geological terms; e.g., in the brachiopods the first high covers four long periods, some 185,000,000 years and the second occurs within a single period, not over 45,000,000 years.

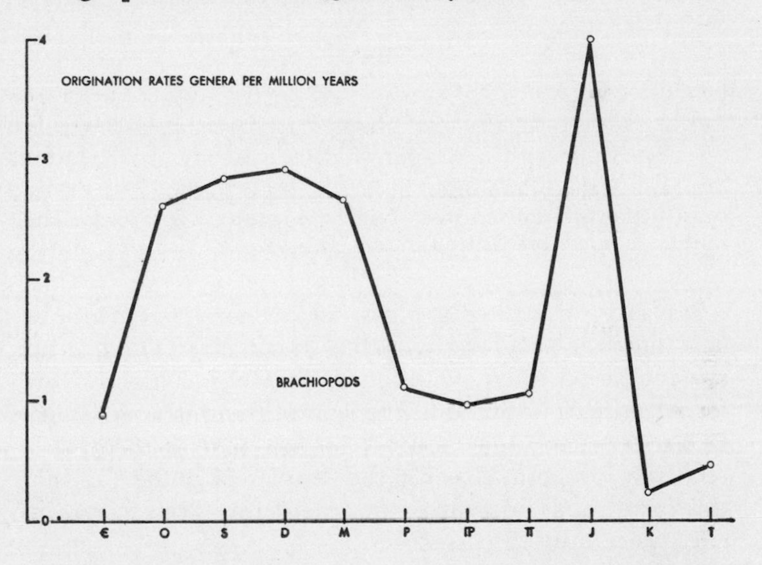

Fig. 20. Rates of evolution in brachiopods. The construction and conventions of the figure are as in figs. 18–19. The time scale runs from Cambrian to Tertiary.

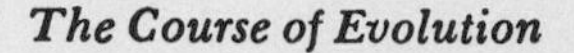

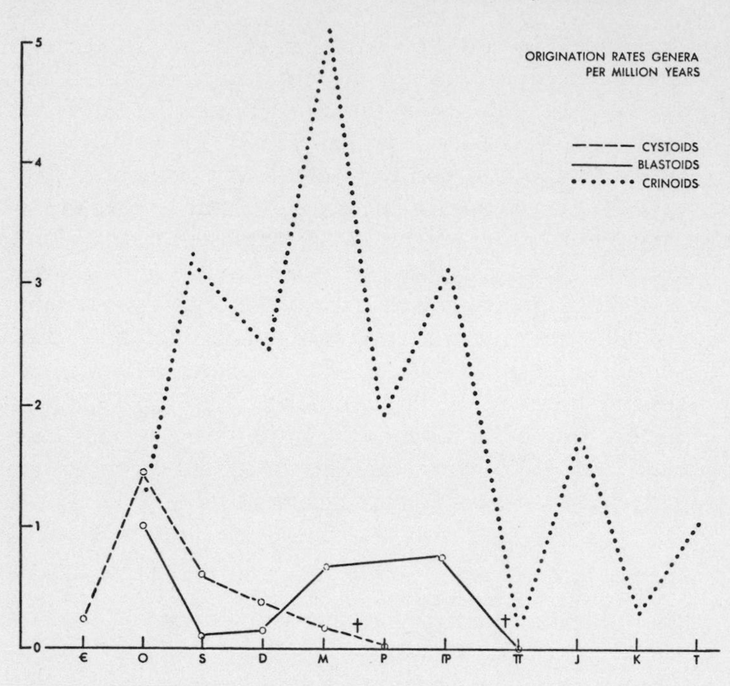

Fig. 21. Rates of evolution in three groups of echinoderms. The construction and conventions of the figure are as in figs. 18–20. The time scale runs from Cambrian to Tertiary.

It is also noticeable that highs and lows for different groups have no clear or strong tendency to coincide. There is no evident period when evolutionary activity was high or low for all animals in general. The nearest thing to a regularity in this respect is the tendency for sea animals to be at a relatively low point either in the Permian or in the Triassic; this corresponds, of course, with the Permo-Triassic marine crisis previously discussed.[4]

4. In his excellent book *Evolution above the Species Level,* previously cited, Rensch has stressed these same characteristics of origination rates and has illustrated them with different examples. He finds some evidence

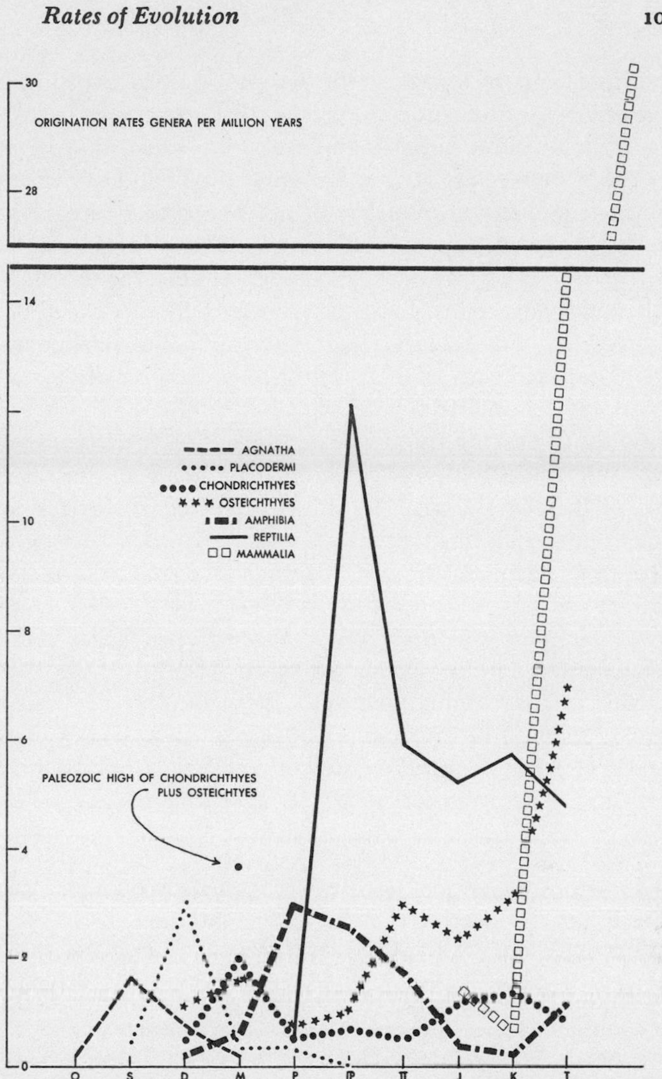

Fig. 22. Rates of evolution in the classes of vertebrates (except birds). The construction and conventions of the figure are as in figs. 18–21. The time scale runs from Ordovician to Tertiary.

This is not to say that the fluctuations in evolutionary activity are quite independent in different groups. They seem to be independent, or nearly so, when comparisons are made between groups in different phyla and of distinctly different habits and habitats. When, however, comparison is made between groups related to each other or of similar ways of life, definite relationships appear. These regularities involve not coincidence but succession of periods of evolutionary activity. For instance among the marine invertebrates called echinoderms, the earliest group, that of the cystoids, rises to a high in Ordovician where two others, blastoids and crinoids, appear at about the same level of activity. (See fig. 18.) Activity in both cystoids and blastoids then declines as that of the crinoids rises to remarkable heights. In the later Paleozoic the blastoids recover to some extent as the crinoids drop to more moderate, but still considerable, levels of evolutionary activity.[5] All echinoderm groups dropped to very low levels of evolutionary activity in the Permo-Triassic crisis.

The rates for the vertebrate classes (fig. 19), irregular as the picture may seem at first sight, show decided relationships of succession. The successive aquatic groups, first Agnatha, then Placodermi, finally Osteichthyes and Chondrichthyes together, go through practically identical phases in the Paleo-

---

of another regularity, a tendency for sea animals (there were no land animals then) to reach a point of high evolutionary activity in the Ordovician. This is not very clearly substantiated by my examples. Rates have not yet been calculated for all known groups of Ordovician animals and it is not yet clear whether generally high evolutionary activity at that time was a real phenomenon of evolution, but it probably was.

5. Another group, that of the echinoids, was also involved and also had a relative high in the late Paleozoic. As regards the Paleozoic, echinoids merely complicate without much changing the picture. Their greatest high was in the later Mesozoic and Tertiary, when crinoids fell relatively low, probably a significant relationship even though it is not a clear example of ecological replacement.

zoic, not simultaneously but each about one geologic period, some 40,000,000 or 60,000,000 years, after the other. Moreover, the peak of each rises somewhat above the last (when Chondrichthyes and Osteichthyes are taken together, as is appropriate in view of their essentially simultaneous origins

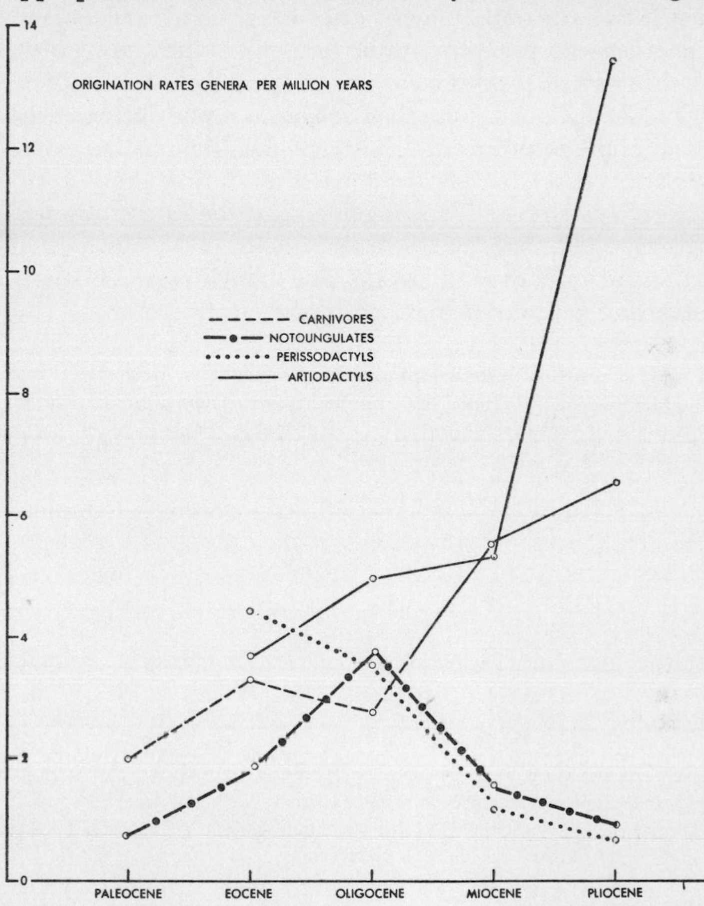

Fig. 23. Rates of evolution in four of the orders of mammals. The construction and conventions of the figure are as in figs. 18–22.

and similar adaptive statuses). Then almost the same sort of successive patterns occur for the terrestrial vertebrates, amphibians, reptiles, and mammals:[6] each rises to a peak and each (except, of course, the last) declines as the next arises. The three peaks are also successively higher, in this case much more markedly so than in the case of the aquatic classes. The times between peaks are, to be sure, somewhat less regular: on the order of 40,000,000–45,000,000 years between Amphibia and Reptilia but more than 200,000,000 years between the peaks of the Reptilia and the Mammalia. Mammalian evolutionary activity has started to decline during the last 10,000,000 years or so. This might suggest the interesting possibility, on the basis of the regularities of past patterns, that the rise of another group is due. The thought is purely speculative and need not be pursued further at this point.

6. The birds, with their radically different habitats, are a special case and one for which the fossil data are inadequate. Their orgination curve, if it could be usefully computed, would probably closely parallel that of the mammals.

IX. FIRST CONCLUSIONS FROM THE RECORD

The record of evolution has now been reviewed, although only in its broader features and in only a few of its many different aspects. We are ready for a first attempt at answering the crucial—if not elegantly put—question, "Just what has evolution been up to?" More profound parts of the answer will be sought later, as we proceed further into the tremendous problem of the meaning of the evolution of life, but here is a good place to pause and to take stock of some of the general features of the evolutionary process that seem to be reflected in the record as a whole.

If you look at the life of the Cambrian and then at that of today, the first and deepest impresison is that of increase. Then there was little or no life on the lands or in the air; now there is hardly a handful of soil or a rod of the earth's surface or a bit of breeze in which life, small or large, does not exist. Even in the seas the increase is richly evident. No fish swam in the Cambrian oceans, no crabs or lobsters moved along their shores, and even the lowly molluscs were then few, simple, and unvaried in comparison with their present exuberance. Yet if you look with speculative eye at the unknown age when life was just beginning, the Cambrian age of life already represents an increase which throws into shade the increase from Cambrian to Recent. Credit is given to Thomas Henry Huxley for an analogy for the filling of the earth with life. He likened it to the filling of a barrel with apples until they heap over the brim. Still there is space into which quantities of pebbles fit before they overflow. Again sand is added, and much of it packs down between the apples and the pebbles. The barrel is not yet full and quarts of water may be poured in before at last the barrel can hold no more.

In the Cambrian many of the apples had been placed in the barrel, although room still remained for more. Addition of pebbles and sand and water had barely started. The beginning of the filling was ancient, for then most of the phyla (we may liken these to the apples) had already appeared, as far as these are susceptible to adequate recording as fossils. We have speculated that this coarser part of the filling took place rather rapidly and ushered in the Cambrian, while the vaster ages of the Precambrian were occupied mainly by the longer processes of the origin of life and its fundamental elaboration to the point where such basic diversification became possible.

None of the basic types, the animal phyla, became extinct. All exist today and most of them in profusion greater than in their early days. Among the lesser types included within them, extinction has touched many, and yet even here the major grades of organization seem to be nearly immortal. Among the vertebrates, for example, even the most ancient, jawless types have their living, indeed thriving, representatives and of the eight great classes only one (the one most dubiously given so high a place in the hierarchy) has disappeared.

No less a personage than Sigmund Freud was bothered by this phenomenon of the survival of the ancient and lowly organismic levels long after they should, he thought, have yielded to the death wish and ceded their places to higher forms.[1] It is unfortunate for both biology and psychology that Freud's busy life had no time for the study of paleontology. His enquiry left him baffled and astray, when paleon-

1. In his interesting book, *Beyond the Pleasure Principle.*

tology could have supplied an answer and Freud's further development of the problem would then surely have enriched the sciences of life.

In the filling of the earth with life, some spaces were filled first, filled well and adequately, leaving neither reason nor possibility for refilling by animals of later development. A protozoan, because ancient and relatively simple, is not therefore an imperfectly adapted type destined for replacement in its own sphere. It is a fully adequate answer to all the problems of life in that particular sphere. The sphere has persisted, and so has the protozoan. The early protozoan was not yet ready to fly the heavens on the wings of a bird, but no more is it possible for the bird to compete with the protozoan for molecular food in the waters of the earth. Each of the phyla and each of the lesser groups that has survived represent basic, permanently sufficient adaptation to a sphere of life, to part of the space in the barrel, from which it has not been dislodged.

New major sorts of organisms have arisen, as a rule, not as more effective followers of ways of life already occupied but as groups extending into and eventually filling new ways. On lower levels of structural types, substitution or replacement of one group by another does occur, as will further be mentioned in this summary, but here, too, expansion into new adaptive zones is the rule of increase for any group. The great and so often repeated phenomenon of adaptive radiation is, for a given group, the working out of this sort of expansion. More universal, but lesser, parts of the broadest processes of increase are the local differentiations and adjustments to conditions at given spots on the earth and for minor variations in its ecology, or the pushing back of merely physical boundaries for one group or another. This is the filling in of the sand and water in the barrel.

To bring Huxley's analogy still closer to a representation of this aspect of evolution, the barrel may be imagined as expansible. Some additions, instead of overflowing it, enlarge it and create new possibilities for further filling. That is, the expansion of some forms of life into new spheres often, in itself, creates other new spheres into which other life may expand. There is not only an increase of life within the possibilities existing for it but also an increase of those possibilities. To a jawless fish in the Ordovician (if we can fancy it endowed with imagination) the idea of land life would have been ridiculous: there was no food on land. The sphere of life for terrestrial animals simply did not exist. It was created only as plants slowly emerged from the waters and clothed the land. As animals followed, living on these early land plants, their emergence created spheres of life for other animals to prey on these. And so the whole cumulative process proceeded, with each new step creating still other possibilities. Even for the oldest phyla new spheres arose in later times. Bacteria and protozoa existed long before vertebrates. The rise of the vertebrates created for them a new possibility, which they have largely exploited: that of life in vertebrate intestines and blood streams.[2]

These invasions of new realms of life and the continual creation of still more new realms, with the packing of each by organisms adapted to all its variations, have resulted in a net increase in the total amount of living matter in the world. Lotka has maintained that this is the main trend of life and the most general law of evolution, in the form of increase in total metabolic processes.[3] The possible implications, for

2. Such developments warrant skepticism as to the opinion of some students that the world is now fully packed with life and that evolution is therefore essentially at an end. Possible ways of life have probably been nearly packed for many millions of years, but the possible ways have kept on increasing. It may be merely lack of imagination not to see that these possibilities can still increase.

3. A. J. Lotka, "The Law of Evolution as a Maximal Principle," *Human Biology, 17* (1945), 167–194.

the welfare of man or his comprehension of life, must be left for later discussion, but here it is to be stated and emphasized that this does appear to be the most nearly universal phenomenon of evolution. It has not, however, been constant or continuous, nor can all even of the broadest features of the history of life be interpreted in terms of this principle alone.

General tendency for increase in the amount of life in the world has been progressive, over long periods, and this suggests problems as to progress in the sense of change for the better. In this sense it hardly follows that later organisms, filling new spheres of life, are therefore more progressive than earlier, which equally filled their own spheres. The whole concept of evolutionary progress is so extraordinarily important for our understanding of the meaning of evolution that it is to be discussed later at some length. Here it suffices to suggest briefly, from the record, the existence of changes other than net increase which are progressive, whether or not they constitute progress in a way related to a human sense of values. We have noted that in the spread of life types of organisms sometimes claim for their own some special place and then for countless years remain static: the horseshoe crab, New Zealand's lizardlike *Sphenodon,* the opossum. Examples are not rare, but they are exceptional. Most groups of organisms have tended to change more notably over any considerable period of time, whether the change be slow or fast.

Groups moving into new places in the economy of nature do not simply occupy it broadly. They radiate into it; they split into distinct types and populations each of which specializes in the occupation of some part of the new sphere. The separate lines of evolution change (not always, but usually) by becoming more narrowly, specifically adapted to some particular way of life, by shifting their adaptations as conditions shift, by developing still other ways of life, or, indeed, simply by what looks like random change with no particular relation to their habits or habitat. Progressive, or perhaps better "successive," change is the rule, with interesting exceptions, whether or not this constitutes progress. Efforts to relate this to some more general principle of progress include such ideas as that of progression from simple to complex, from more to less dependence on environmental conditions, and the like. So far as they are important to our theme, these will be reviewed later. The grand lines of the objective record of life do not seem to impose any such ideas on us or to make such principles obvious, even though they may be real and effective in some less apparent way. Whether Recent man is to be considered more complex or more independent than a Cambrian trilobite will be found subject to qualifications and dependent on definitions.

Previous examples have sufficiently displayed another widespread phenomenon of the record: that of replacement. The phenomenon does not have an obvious or simple, although it may have a more subtle, relationship to over-all increase in life, for it may (as far as we can tell) amount simply to substitution of one organism by another no more effective in the increase of mass or energy of life. That one form replaces another *because* it does contribute to such increase is, by and large, more an expression of faith than of demonstrated fact. It may be, indeed probably is, true of certain cases but not of all. This negative seems to be proved in the record by the instances, which are rather numerous, of deferred replacement. Porpoises and dolphins, which are mammals, have rather closely replaced ichthyosaurs, which were reptiles. But ichthyosaurs became extinct millions of years before porpoises and dolphins arose, and during the interval this adaptive zone was simply empty. Of course there are the possibilities that the zone, the opportunity of such a way of life, ceased to exist in that interval or that some other type filled it then, but it does not look so on the face of the evidence. Nor do such explanations satisfy for all other examples, bats replacing pterodactyls, various land mammals replacing dinosaurs, and many others. The refilling of the zones in such cases represents expansion in accord with the principle of increase, but its aspect as a replacement of older forms in the zones is hardly reducible to such a principle.

It is noteworthy that such replacement is usually an approximate, not an exact, duplication of earlier ways of life. Conditions change and groups of later origin are never quite like those of earlier ages. Bats are only broadly, and not closely, similar in habits to some, and not to all, pterodactyls. This failure truly or fully to reoccupy a zone which has itself changed is especially characteristic of delayed replacement, but an aspect of replacement is nevertheless present.

Many more instances of replacement well exemplified in the fossil record evidently do represent the ceding of one group to a more successful competitor, whatever the basis of the success. Thus we have seen examples of the members of an early radiation yielding to the expansion, or radiation anew, of some "progressive" type that itself arose also from that radiation: witness the creodonts and the fissiped carnivores. Or one of two parallel groups may be reduced or partially or wholly replaced by the expansion of the other, as the bony fishes expanded from fresh to salt water to the detriment of the sharklike fishes in the latter. Or again, a radical radiation may produce types convergent toward some already existing and somehow fit to replace the latter, or at least to make place in their environment, as the aquatic mammals and reptiles have done in seas already swarming with fish of all sorts.

As the trend toward increase is a principle of addition or multiplication and the common process of replacement is a principle of substitution, so, in continued mathematical terms, extinction represents a principle of subtraction or division. The causes of extinction and its implications have such significance for philosophical understanding of life that, again, the subject will require further attention in a later section. Its interpretation is by no means obvious on the face of the record where, nevertheless, the fact of extinction is striking enough. There are, in the main, three repeatedly recorded sorts of extinction. In one, a group of organisms simply changes to such a degree that it is no longer the same thing. The ancestral form is no longer there, so it is extinct, and yet this is not extinction in a truer sense, for the group lives on in different form. Other cases are those of complete replacement, when one group occupies the sphere of life of another and the latter disappears or terminates in consequence. ("In consequence" is a conclusion not forced or proved by the record, but sufficiently justified in most cases.) Finally and most baffling of all are apparent cases of termination without replacement, an apparent negation of the principle of increase, leaving a void in the expanding mass of life. Such cases, too, appear in the record and only rather speculative interpretation can seek to do away with the paradox that they present.

The processes of addition, substitution, and subtraction add up to the total picture of rise and fall of the various groups of organisms such as has been exemplified at considerable length in previous pages. We sought there, and failed to find, any sort of standard pattern of increase and decrease or of acceleration and deceleration of evolutionary rates. Though the total may suggest a slow rise, fluctuating, to be sure, to a single maximum, no such regularity is evident throughout the groups that go to make up the total. Alternative patterns that might have seemed a priori probable (especially under some particular theories as to the cause of evolution) include sudden expansion and long decline, or steady rise to a climax and equivalent decline from it. These patterns occur, but neither is universal nor even so predominant as to be called usual. Decidedly uncommon is geologically long continuance of any group at a fairly constant level of abundance or of evolutionary rate.

Climatic, geographic, and other physical influences on the evolutionary record have been exemplified but have not been stressed. This is not because they lack importance but because most of their effects have evidently been local, erratic, and confined to particular groups. The very lack of any roughly standard pattern for the histories of groups of animals shows that there has been no over-all trend of physical influences on them and that completely general principles affecting life as a whole are not to be sought here. From this broadest point of view the earth as an abode of life has been more or less the same since the Cambrian, at least, except as life itself has occupied and has changed the different parts of the physical environment. Yet the local and temporary fluctuations in physical conditions have often had important and enlightening effects on the particular histories of the separate groups of animals, as sufficiently exemplified, for present purposes, by the Permo-Triassic crisis for many marine organisms or the history of isolation and reunion of South America and the effects of this on its land life.

Perhaps at the end the strongest effect left by the record of life will be one of an odd randomness in it, a sense that it is dominated by a sort of insensate opportunism. Such opportunism is not contradicted by the most nearly general features of the record: the "barrel-filling" effect, or the principle of total increase (with the concomitant tendencies to expand the barrel and to refill when emptied at any point). On the contrary, the lack of fixed plan in detail but the tendency to spread and fill whenever possible is exactly such a picture as would result from the impulses of a random opportunism.

Yet not all is random in this record. We know (or let us

grant such knowledge at this point) that change must have direction and cause. And we sense a need to evaluate changes and their causes. From this relatively superficial examination of the record of life let us pass to an attempt to interpret it more profoundly, and first to the question as to what sorts of causes may be operating in this tangled fabric of life.

PART TWO

The Interpretation of Evolution

"To be content with the religious answer—always apt to become a soft pillow to the easy-going—is to abandon the scientific problem as insoluble, and there can be no greater impiety than that. It is surrendering our birthright—not for a mess of potage, it is true, but for peace of mind. Therefore man is true to himself when he presses home the question: How has this marvellous system of Animate Nature come to be as it is?"

J. Arthur Thomson, in the first Terry Lectures, *Concerning Evolution.*

X. THE PROBLEM OF PROBLEMS

What forces have been acting throughout the history of life? This is the problem of problems for evolution and for life itself.

Are these the same forces that act throughout the material universe, different in their results only to the extent that the matter on which they are acting is differently organized? Or do they include forces peculiar to and inherent in life, essentially different from the mechanistic forces of cause and effect in the purely material realm? Or, again, do they involve some principle that transcends both matter and life itself, a force that brings about progression toward foreordained goals, and that not only negates but also reverses materialistic cause and effect so that effect precedes cause?

Each of these questions has been answered affirmatively by one or another scientist or philosopher. They embody the three main possibilities in the solution of the problem of problems and each has its ardent supporters. The first solution is that of the philosophically naturalist, materialist, mechanist, or causalist—the terms are not synonymous but in this connection they designate roughly congruous philosophical points of view. The second solution is that of the vitalist, the student who maintains basic distinction between the principles and forces reflected in vital and those reflected in material phenomena. Adherents of the third solution are almost always also vitalists in this sense (although in fact there is no logical necessity for them to be such), and this solution has commonly been considered no more than another of the innumerable varieties of vitalism. It adds, however, an alternative that differs from vitalism alone as profoundly as vitalism does from materialism. If life, as the vitalists submit, has its own, nonmaterial forces, it does not necessarily follow that these tend toward a final goal or achieve a transcendental purpose. That life, or that the whole cosmos, does have such a finality is the view of the finalist.

There has been a great deal of misunderstanding and name calling among the materialists, vitalists, and finalists. The vitalists and finalists usually impute to the materialists the views that there is nothing in the universe but pure mechanism and that there is no essential difference between life and nonlife. Some materialists (in a strict sense) have accepted these imputations and have attempted to defend these propositions. Their purely or merely mechanistic view was more popular a generation or two ago, arising in the first enthusiasm over the great nineteenth-century discoveries in science, than it is today, but it still has able supporters. Others have found this viewpoint indefensible, scientifically or philosophically, and have decided that they were, after all, vitalists rather than materialists, vitalists of some special sort, adherents of "emergent vitalism" or "monistic vitalism."

Now it is merely silly to maintain that there is no essential difference between life and nonlife. The vitalist is naïve indeed who proclaims the discovery of such a difference and concludes that because living organisms operate otherwise than if they were not alive, therefore materialism or naturalism is wrong and life is a new element or a vital principle or what not. This, the common argument against materialism as it is defined by its opponents, is analogous to claiming that there is a "fire substance" or "fire principle" because fire has properties and phenomena peculiar to it and absent in nonfire. This, indeed, was the theory of our ancestors who gave the name "phlogiston" to the supposed fire element or principle,[1] very much in the same way that some vitalists have given names such as "entelechy" to a hypothetical life element or principle. We know that fire is not such a separate element or principle but that it is a process and organization of matter in which the behavior of matter is different from that in nonfire. Similarly, the materialistic view is not abandoned when life is seen as a process and organization in which the behavior of matter is different from that in its nonliving state.

Granting, as any reasonable person must, that there is an important difference between life and nonlife, you may, if you wish, call the different behavior of matter in life "vitalistic," but this accomplishes nothing and means nothing that was not already obvious. It is an example of the naming fallacy to call this an explanation or a contribution to evolutionary theory. The real issue between materialists and vitalists is not whether life has its own principles and functions, a proposition admitted by most modern materialists and (it seems to me) self-evident. This sort of "emer-

1. This fascinating aberration and the long struggle to correct it are worthy of close study by anyone interested in problems of truth and error. See any good history of science, such as A. Wolf, *A History of Science, Technology, and Philosophy in the Eighteenth Century* (New York, Macmillan, 1939).

gent" as opposed to "substantive" vitalism[2] is embraced within most of the less crude brands of materialism. A valid issue arises only as regards substantive or dualistic vitalism, the view that there is a life substance or principle which is independent of and in addition to the substances found in inorganic nature and the laws of their behavior.[3] On the other hand, the distinctive claim of materialsm is not stupid denial of special attributes to life, but the view that the substances and the principles involved in organic evolution are those universal in the material world and that the distinctive attributes and activities of life are inherent in its organization only.

This clarification is essential in the marshaling of evidence that may bear, one way or another, on choice between materialistic and vitalistic interpretation of the history of life. Beyond this are further philosophical implications, to be discussed more fully in the last section of this book but to be mentioned here in passing to forestall possible misapprehension. The view of life as mere mechanism or automatism represents one extreme of the general field of materialism, an extreme now rather widely abandoned by materialists. It is compatible with materialism, which in this form is less ambiguously called naturalism, to hold that life and the

2. See, for example, C. D. Broad, *The Mind and Its Place in Nature* (New York, Harcourt, Brace, 1925).

3. There is still a third among the shades and varieties of vitalism, one that holds that a vital principle, some form of "mind," "mentation," or "psychism," is inherent in matter and present as well in things we call nonliving (even an atom or inorganic molecule, for example) as in those we call living. Although I think few adherents of his cult realize it, that was part of the idiosyncratic theology of the late P. Teilhard de Chardin, as expressed for example in *The Phenomenon of Man* (New York, Harper and Brothers, 1959). Teilhard's views on evolution were also extremely finalistic. The current passionate development of a Teilhardian cult shows how strong an emotional grip such nonscientific views still can have.

universe involve more than objective material and mechanism. At one end of the enquiry the origin and nature of the existing materials and mechanisms have no evident materialistic explanation even though, given these as existing, their operation may be purely naturalistic. At the other end the possibility is not excluded that, within the unique organization of matter that is life, these operations may develop choice, values, and moral judgment. The existence of these qualities is also a basic tenet of most vitalist theories, which tend, however, to maintain that they are inherent in, rather than developed by, life.

The distinctive finalist belief is that of progression toward a goal or end. The end is not reached, the finalist believes, because of what goes before, but what goes before is but a means for reaching the end. The end, although later in time, is, then, the cause and the preceding course of history is the effect. The history of life is thus to be viewed as purposeful, and (it almost goes without saying) finalists usually consider man as the essential feature of that purpose. Such a view is not necessarily inconsistent with naturalism: the forces of history could be materialistic and yet have been instituted as a means for reaching an end. Finalists more commonly, however, hold subsidiary opinions that are vitalistic rather than materialistic. The purposefulness of evolution is considered inherent in, or special to, life, or an essential vital essence is seen as a means of reaching the goal.

Scientists and particularly the professional students of evolution are often accused of a bias toward mechanism or materialism,[4] even though believers in vitalism and in final-

4. See for instance E. S. Russell's witty and persuasive but oddly inconclusive book, *The Directiveness of Organic Activities* (Cambridge, England, University Press, 1945). Russell ridicules the mechanistic bias of scientists and develops a finalist thesis on the basis of the directiveness or purposiveness of life activities which are said not to be reducible to "activities of a lower order," i.e., nonvital. Essentially the same views

ism are not lacking among them. Such bias as may exist is inherent in the method of science. The most successful scientific investigation has generally involved treating phenomena *as if* they were purely materialistic or naturalistic, rejecting any metaphysical or transcendental hypothesis as long as a natural hypothesis seems possible. The method works. The restriction is necessary because science is confined to material means of investigation and so it would stultify its own efforts to postulate that its subject is not material and so not susceptible to its methods. Yet few scientists would maintain that the required restrictions of their methods necessarily delimit all truth or that the materialistic nature of their hypotheses imposes materialism on the universe.

Although the metaphysical cannot be directly investigated by the methods of science, it results may be. It is, indeed, one of the greatest values of our present subject that it can serve as a means of testing the fundamental philosophies of naturalism, vitalism, and finalism. Vitalism and finalism involve elements postulated as beyond the reach of purely material scientific investigation. Yet the truth of these philosophies would involve material consequences in the history of life. The investigation of these possible consequences is within the scope of scientific method, which therefore can provide evidence on which to base a choice among naturalism, vitalism, and finalism, as well as among the variety of particular theories that have been elaborated within the framework of each of these philosophies. Conviction as to the essential truth of materialism and naturalism need not, then, be the

---

have been developed at great length in America by E. W. Sinnott in three books, the third of which is *The Biology of the Spirit* (New York, Viking, 1955). This finalistic argument is also embraced, with much else, in the Teilhardian canon, mentioned in the last footnote. I have discussed these personal forms of vitalism and, as I believe, refuted them in Chapter 11 of *The View of Life* (New York, Harcourt, Brace and World, 1964).

result of bias or of the limitations of the scientific method, but may be the result of careful evaluation of evidence. A scientific bias cheerfully confessed is the belief that the results of such evaluation are likely to be nearer the truth than are the inclinations of personal preference or the traditions of a less knowledgeable and more gullible past.

Such phenomena of evolution as should provide particular crucial evidence of this sort are now to be especially considered. The field of evolution in all its various phenomena and processes is so vast that selection must be exercised. An acceptable general theory of evolution must be prepared to explain all these phenomena, or at least to demonstrate that such explanation is a reasonable probability within its framework. All, however, are not equally crucial for the evaluation of the proposed theories. Certain aspects of evolution are evidently particularly likely to reveal possible underlying general causes and to help to discriminate among these as to their reality or probability. In the long quarrel between materialists and vitalists it has, for instance, been recognized on both sides that the phenomenon of orientation in evolution—of the real or apparent tendency for evolution to proceed along limited, directional paths—is crucial for any eventual decision between the schools. It is such lines of evidence as this that require particularly critical discussion now.

It will hardly be supposed that all materialists, all vitalists, or all finalists have been in agreement among themselves. On the contrary, within each broad school of thought the most diverse shades of opinion have existed, and two general theories as to the causes of evolution may both be, for instance, thoroughly naturalistic and yet stand in strong contradiction to each other. Only less important than choice among the broader philosophies is critical evaluation of these many theories. Evidence for such evaluation will also be sought in dealing with processes of evolution in the following pages, and a historical summary and attempted judgment of the more important of these theories will follow.[5]

5. The complex of causes and mechanisms of evolution expounded in Chapters XI–XV is now commonly called the synthetic theory, "synthetic," as explained in Chapter XVI, because it is a synthesis of data, references, and conclusions by numerous researchers in all fields of the life sciences. It may be helpful at this point to gather citations to some of the more important, fairly recent books in English on the general theory of evolution, even though some are also repeatedly cited elsewhere:

A. J. Cain, *Animal Species and Their Evolution* (London, Hutchinson, 1954). Th. Dobzhansky, *Genetics and the Origin of Species* (New York, Columbia Univ. Press, 1951), and *Evolution, Genetics, and Man* (New York, Wiley, 1955). V. Grant, *The Origin of Adaptations* (New York, Columbia Univ. Press, 1963). J. S. Huxley, *Evolution the Modern Synthesis* (London, George Allen and Unwin, 1963). J. S. Huxley, A. C. Hardy, and E. B. Ford, editors, *Evolution as a Process* (London, Allen and Unwin, 1954). E. Mayr, *Animal Species and Evolution* (Cambridge, Harvard Univ. Press, 1963). B. Rensch, *Evolution above the Species Level* (New York, Columbia Univ. Press, 1960). A. Roe and G. G. Simpson, editors, *Behavior and Evolution* (New Haven, Yale Univ. Press, 1958). I. I. Schmalhausen, *Factors of Evolution* (Philadelphia, Blakiston, 1949). G. G. Simpson, *The Major Features of Evolution* (New York, Columbia Univ. Press, 1953), and *This View of Life* (New York, Harcourt, Brace and World, 1965). G. L. Stebbins, Jr., *Variation and Evolution in Plants* (New York, Columbia Univ. Press, 1950). S. Tax, editor, *Evolution after Darwin* (Chicago, Univ. of Chicago Press, 1960).

XI. ORIENTED EVOLUTION: ORTHOGENESIS AND TRENDS

A widely held opinion regarding the course of evolution has been that it tends to occur in straight lines, that once a certain sort of structural change has started in a given group this tends to continue without deviation indefinitely or to the extinction of that group. The example most often given is that of the change from eohippus to the modern horse, *Equus,* represented as involving a gradual increase in size, reduction in number of toes, increase in height and complication of teeth, and generally a transformation of all characters of eohippus to those of *Equus* in a steady, undeviating way. It happens that this usual example has been badly misrepresented. Things did not, as a matter of observable fact, occur in just that way in the history of the horse. Yet the fact remains that the sequence cannot be considered random. The changes involved do have direction and orientation, even though these were not as regular as they have usually been represented. And so, in hundreds or thousands of other cases, it seems clear that there is an orientation of some sort.

The record as a whole gave an effect of opportunism, of divergence into nearly all available ways of life, yet this in itself is an oriented rather than a random process. Examined in greater detail, the record seems still less to be purely random, and descent within any one line usually gives the impression of being rather stringently oriented in many, although commonly not in all, of its features. There are, also, sorts of structural changes so widespread in different lines that they seem to represent general tendencies, a common orientation, or so-called "laws" of evolution. (Like human laws, the "laws" of evolution are often broken and, indeed, they are human constructions which we impose on the complex phenomena of nature and which nature is under no obligation to follow.) Among these, one of the best substantiated is a tendency for increase in size. Little eohippus and big *Equus* provide, again, a familiar example. Others may be found in almost any group of animals: medium-sized Eocene and Oligocene proboscideans and their enormous Pleistocene and Recent successors among mastodons, mammoths, and elephants; relatively little Triassic dinosaurs and the gargantuan Jurassic and Cretaceous forms; small Cambrian sea snails working up to forms six inches or more in height in the early Carboniferous and to truly enormous sizes in later and Recent times—in familiar example, the conchs so abundant off the Florida keys.

The widespread occurrence of oriented evolution being established, questions arise as to just how characteristic it is, whether universal, usual, or only occasional, the existence and nature of exceptions, the relationship of oriented evolution to environmental factors, and other enquiries leading up to the question as to what orienting force or forces are involved. On this last point the main alternatives are whether orientation, or change in a straight line, is something inherent in life and its evolution, a hypothesis held in common by most vitalists and finalists; or whether it is dependent only on the physical possibilities of the situation and on the interplay of organism and environment, the usual materialist hypothesis. In the first case, orientation should be universal or nearly so except, perhaps, when gross environmental fluctuations make it quite impossible. It should be not merely directional but unidirectional, that is, not merely tending in some direction but long maintaining a single direction. Above all, it should not fluctuate according to the adaptive advantage to the animals concerned and there should be instances when the long trend in one direction is inadaptive finally or throughout, that is, is harmful rather than useful to the harmonious functioning of the animals in their environments. On the other hand, under the various more naturalistic hypotheses, orientation may be expected to be less rigid, to have many exceptions, to allow numerous changes and even occasional reversals in direction of evolution, and to be related, on the whole, to adaptive advantage for the animals concerned.

It is easy enough, in the first place, to observe that evolution along straight lines, or even the rigid orientation of evolution in a less simple way, is far from universal. It is only a tendency with so many exceptions as hardly to constitute a rule.

Differences in rate, kind, and direction of evolutionary changes are strikingly characteristic of the true facts in the prime example constantly cited as supposed proof that evolution occurs without such differences, in an orthogenetic[1] way: the horse family. Consider the main trends in evolution of horse teeth, for instance. One type of change in them was a tendency for premolars and molars to become more alike—all but the first premolar—with the result that six cheek teeth, three premolars and three molars, have come to have almost the same pattern. This went on rather steadily during

1. Evolution in a straight line is commonly called "orthogenesis" and will be found discussed under that title in most books on evolution. The term has, however, been kicked around so much that hardly any two students mean exactly the same thing when they use it. To some it means little more than that evolution is not completely random. To others, use of the term implies granting the whole finalist thesis of undeviating progress toward a goal. Arguments over orthogenesis are unduly obscured and complicated by entirely unimportant semantic difficulties. It usually is understood, however, to mean a postulated inner urge or inherent tendency for evolution to continue in a given direction.

the Eocene, reached essential completion at the end of that epoch, and then stopped. Another change was complication and, in other ways, alteration in the pattern of the molars (and of the premolars, as they became more like molars). This went on in a fluctuating way but in a single line from eohippus through the Eocene and Oligocene—in a single line only to the extent that the evolving group in America, which eventually gave rise to *Equus,* was then a single line of major populations. Other lines from eohippus occurred in Europe, and even in the Eocene these gave rise to diverse lines with quite different tooth patterns. Then in the Miocene the hitherto more nearly unified American line of the horse family split into at least five and probably more major branches, contemporaneous genera, *each* of which had a distinct and characteristic molar pattern. Some of these tended to have simpler patterns than their ancestry, and some more complicated. One of these again split into at least five distinct lineages and again each developed a different sort of pattern, varying but now all more or less comparable in degree of complication. In height of teeth, too, the various lines in the Miocene and Pliocene differed radically. In some the teeth remained low, increasing only in correlation with the sizes of the animals. One Miocene line rapidly developed exceptionally high teeth. This was not a continuation of the "orthogenetic" trend of the Eocene and Oligocene horses but a new direction of evolution. Although there is, in fact, no strictly straight line of horse evolution, the nearest thing to one leads not from eohippus to *Equus* but from eohippus to an extinct group, decidedly different from the living horses, called *Hypohippus.* (See fig. 24.)

Every feature of horse evolution tells a comparably complex story if this is examined in detail and in all the divergent lines of the horse family. The feet, to supplement the example here, hardly evolved at all in the Eocene, then evolved rapidly to a basic three-toed Oligocene type which remained nearly static in some later lines, in others evolved gradually to different three-toed types mechanically sounder

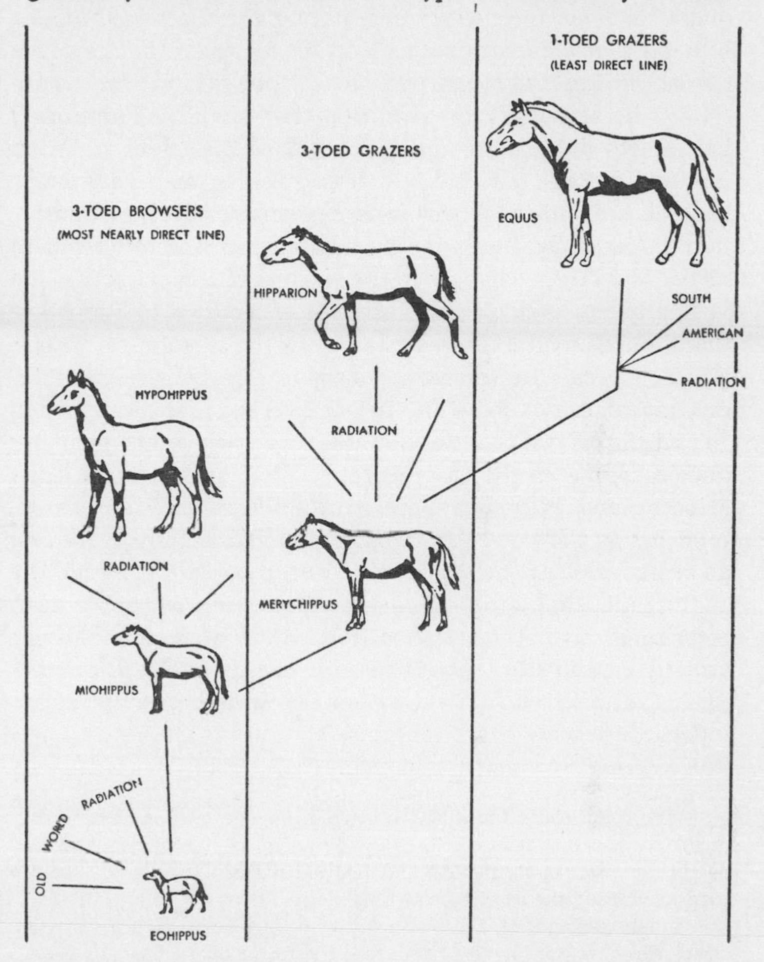

Fig. 24. The evolution of the horse family. The animals shown are only a few of many in the family, and lines of evolution were more numerous and complex than here indicated in detail. The various forms are drawn to the same scale.

for larger animals, and in one line, only, finally evolved rapidly in one phase to a one-toed type. This, again, did not really continue the usual trend among the three-toed types but was a new evolutionary direction for them. In the particular lineage from eohippus to *Equus,* general foot mechanics became first more complex, then simpler. The number of toes did not change at even pace from four (in the forefoot) to one, but changed in two spurts, first from four to three, then much later from three to one, each rapid transition followed by slower mechanical adjustment to the new sort of foot and to changes in the weights of the animals.[2]

The horses even provide us with exceptions to the rule that animals tend to increase in size in their evolution. During the Eocene the record, contrary to a rather general impression, does not show any net or average increase in size. In fact the known late Eocene horses average rather smaller than eohippus in the early Eocene. Then still later, in the Miocene and Pliocene, there were at least three different branches of the horse family characterized by miniature or decreased size (*Archaeohippus, Nannippus, Calippus*), while at the same time other lines were, according to "rule," increasing in size. At that time, too, others were fluctuating around a mean size without notable change and still others developed different species of decidedly different sizes—as, indeed, is the case in *Equus* today.

2. Details of these and other changes in horse evolution as it really occurred are rather fully set forth in the technical literature. Unfortunately, however, more general and popularized accounts have usually tried to simplify the story by omitting the real diversity of horse evolution and discussing it *as if* it were a case of orthogenesis. Even paleontologists who have not handled the actual specimens or digested the great volume of very technical publications have sometimes been seriously misled by this spurious simplification. I have elsewhere tried to give a less misleading nontechnical account of this crucial example of evolution in action in *Horses* (Garden City, Anchor Books [Doubleday], 1961).

There is increasing evidence that mammals in general, especially some of the relatively large forms, have tended to decrease in average size since the Pleistocene ice age.[3] In itself this negates any invariability in the rule of increase of size, and it certainly strongly suggests adaptive response to climatic conditions as opposed to size control by some inner tendency or life urge within the organisms alone. We know that climates have tended to become warmer since the Pleistocene. We also know that closely related living mammals show the adaptive phenomenon of being, on an average, relatively smaller in warmer climates. It is certainly reasonable to suppose that this is the same sort of phenomenon involved in size decrease from Pleistocene to Recent.

In this connection, it is known that many large animals of the past became extinct and are not the ancestors of their smaller living relatives. Mammoths were not ancestral to smaller elephants. (As a matter of fact, most mammoths were no larger than some living elephants, but a few were.) The elephantine ground sloths were not ancestral to the little living tree sloths. The dinosaurs were not the ancestors of the small lizards of later times. But this does not mean that forms that *were* the ancestors of living animals were not also somewhat larger than the latter at one time or another, and such does appear to be the case for some of them.

Some paleontologists have been so impressed by the frequent trend for animals to become larger as time goes on that they have tried to work it the other way around. If they

3. The evidence has not all been compiled in one place but it is practically conclusive for some groups and strongly suggestive of a general trend. See, for instance, C. B. Schultz, and W. D. Frankforter, "The Geologic History of the Bison in the Great Plains" (*Bull. Univ. Nebraska State Mus., 3* [1946], 1–10); and D. A. Hooijer, "Pleistocene Remains of *Panthera tigris* (Linnaeus) Subspecies from Wanhsien, Szechwan, China, Compared with Fossil and Recent Tigers from Other Localities" (*Amer. Mus. Novitates,* No. 1346 [1947]).

find, say, a Pleistocene bison that is somewhat larger than a Recent bison (so-called *Bison taylori,* associate and prey of early man in America, is a good example), then they conclude that it is not ancestral to later bison *because* it is larger. You can establish any "rule" you like if you start with the rule and then interpret the evidence accordingly.

This suggests some brief digressive but pertinent remarks on certain psychological factors in the interpretation of the necessarily imperfect evolutionary record. A really complete evolutionary lineage is never preserved; this would mean fossilization and recovery of all the animals that ever lived as parts of that lineage, an absurdly impossible chance. Museum curators know that even if the miracle of fossilization occurred, recovery and study would be impossible if only on grounds of expense. The actual data, then, normally consist of relatively small samples of the lineage, scattered more or less at random in space and time. The process of interpretation consists of connecting these samples in a way necessarily more or less subjective, and students may use the same data to "prove" diametrically opposed theories. Thus with data like those schematically shown in figure 25A, one student may believe that evolution is always or basically orthogenetic. He will then conclude that the samples represent variations around a series of different straight lines of evolution, perhaps from four different lineages, with the earlier parts unknown for the later lines (figure 25B). Having composed this figure, he will adduce it as evidence that evolution is orthogenetic.[4]

Another student may hold the opinion that evolution is a discontinuous process, consisting of leaps from one level or group to another, without intermediate forms. He will

4. The example is diagrammatic but not imaginary. See, for instance, the interpretations of lineages of mastodons and elephants in H. F. Osborn's *Proboscidea* (New York, American Museum of Natural History, 2 vols.: 1936, 1942).

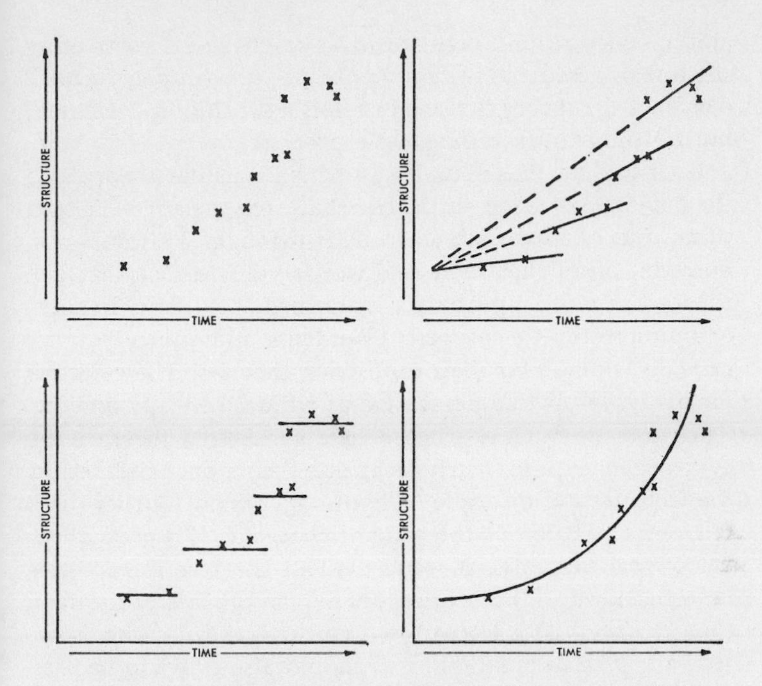

Fig. 25. A hypothetical example of known facts of animal history and various ways of interpreting them. A (upper left), observed facts. B (upper right), interpretation as a series of separate orthogenetic lines. C (lower left), interpretation as a sequence of stages with each arising by one leap, without transition. D (lower right), interpretation as a single line of evolution, changing in direction.

then interpret these samples more or less as shown in figure 25C, and in turn will claim that this record is evidence for his theory.[5]

Still another student will notice that the over-all pattern of these samples is, in fact, oriented or correlated in time

5. Closely analogous examples of subjective arrangement of evidence to favor this type of theory may be found, for instance, in O. H. Schindewolf, *Grundfragen der Paläontologie* (Stuttgart, Schweizerbart, 1950).

and in structure, although not in a straight line. Interpreting the line of descent as in figure 25D, he will consider this evidence for the theory that evolution is a continuous, oriented, but not inherently orthogenetic process.

Each student thus actually puts his particular theory into the data, and it is not surprising that each then gets his own theory out of these data when he is through. This does not, however, mean that such evidence is valueless, or that one theory is as good, as probable, as another. It explains diversity of opinion among competent students and shows how all can cite evidence for their conflicting theories. The evidence for his views may be persuasive as adduced by any one student, and yet when all theories are considered together the evidence may quite clearly favor one theory over another. In the hypothetical example of figure 25, which I believe is a fair representation of the general character of a great number of real examples, it seems evident that the interpretive pattern shown in D is more probable, on the face of the data, than B and C. D is less subjective: the trend it shows is objectively present even if B or C should also prove to be present. It is the most probable inference on sound statistical principles. It, alone of the three, follows the famous maxim of William of Occam, fundamental in logic and in scientific investigation, that entities (in theories or hypotheses) are not to be multiplied unnecessarily.

Although in cases like this example any of the conflicting theories *may* be true, there are other cases in which one of them *must,* within reason, be true. It does not wholly follow that the same theory *must* also be true in all cases where others *may* be, but a strong probability is certainly established. Such decisive examples do exist for our present subject. The horse family is one. Incomplete as they are, the data are so abundant and, aside from one or two minor points, so completely continuous that it is fair to say that they *cannot* be fitted to an interpretation of type B or C in figure 25. They certainly represent a sequence like D of that figure, complicated only by the presence of repeated branching which is fully consistent with the theory symbolized by D and largely inconsistent with those of B or C.

It is easy enough to see that, in evolution as a whole, departure from a straight-line pattern must have been exceedingly frequent. Every rise of a distinctively new adaptive type among organisms *necessarily* represents a change in direction of evolution. A fish evolving in a constant direction remains a fish and does not become an amphibian. Whales, which certainly arose from quadrupedal land mammals, cannot by any juggling of the data be represented as arising by continuation of the trend that gave rise to quadrupedal mammals. Bipedal man must equally represent a new trend, a decided change in evolutionary direction from that giving rise to his quadrupedal ancestors.

Similarly each of the many thousands of major and minor adaptive types arising in the evolution of life represents a change, large or small, in the direction of evolution. Strictly straight-line evolution is confined to relatively minor progressions within established types. Even here this phenomenon seems less common than claimed by those who impose an orthogenetic pattern on any sequence too imperfect to exclude this possibility. Adequately known sequences rarely show this sort of purposeful-looking progression straight toward an end which might be a goal. The most nearly direct evolutionary lines seem almost always to be short, geologically speaking. And any curve seems nearly straight if only a short segment of it is seen.

The extreme view that evolution is basically or over-all an orthogenetic process is evidence that some scientists' minds tend to move in straight lines,[6] not that evolution

6. In a session on orthogenesis at an international conference on evolutionary problems held in Paris, an eminent student startled his colleagues by proclaiming that evolution consists of nothing but millions

does. The grand fact is that rigid orientation of evolutionary change in a single direction is not inherent in the evolutionary process or even particularly characteristic of it, and yet that the process is not fully random. Orientation is usual, but the directions diverge for different groups, and for single lineages the direction may, so to speak, twist and turn in a remarkable way. *Something* operates in evolution to keep changes, for greater or less periods, progressive within delimited paths, but these paths may not be straight and may have unexpected ends.

There are, of course, certain limitations always present in any existing situation. Evolution fully at random in an unlimited, or even in a very large, number of directions is never possible to a group of organisms. Changes in size of whole organisms or of various of their parts are probably the commonest sort of evolutionary change, and such evolution can be in only one of two directions: toward larger or smaller. There is no conceivable alternative. More subtly, given a metabolic system that requires intake of oxygen, the evolution of a group of animals is limited to environments where oxygen is present and to such structural changes as retain or provide apparatus for extracting oxygen in that environment. Without change of environment, a deep sea fish cannot evolve lungs nor an eagle gills. Profound limitations on possible directions of evolution for any group of organisms are imposed by two indisputable factors universally present: surviving organisms must meet the minimum requirements of life in an available environment, and changes can only occur on the basis of what already exists. The environments available are not limitless, and each permits survival on its

---

of orthogenetic lines. This is the most extreme case of orthomania, or straight lines before the eyes, known to me, but milder cases of this affliction are farily common. "Paléontologie" (*Colloques Internationaux du Centre National de la Recherche Scientifique,* Paris, No. XXI, 1950).

own terms only. At any time the possibilities for future change within a given group are also always limited by what the organisms have already become; their past is one of the determinants of their future. Evolution is amazingly versatile in adapting the materials at hand to other uses, as will be seen later on, but its possibilities are always limited by the nature of those materials and by the external requirements for survival.

Even these broad, rather obvious limitations suggest strongly that orientation in evolution is not determined solely by some characteristic within the evolving organisms or solely by external factors in their environment, but by both and by interplay between the two.

It is an old theory, probably the first theory of evolution to be proposed, that changes in groups of organisms are not merely limited but also directly induced by changes in the environment. If this were true, then the environment alone would be the orienting factor in evolution, the nonrandom or antichance factor that clearly does exist somewhere in the process. The theory finds support in the universal integration of organisms into their environments, the evident close correspondence between most of their structures and activities and the needs imposed by their places in nature. It is clear enough that environment must have something to do with the orientation of evolution, but the idea that it is the direct or sole *cause* of oriented change has broken down and has had to be abandoned by almost all students of evolution.[7] In the wide picture of life on the earth the evidence

7. "*Almost* all" does not really mean that the theory needs still to be very seriously considered. So diverse and erratic are the operations of the human mind that no theory is likely completely to lack supporters. After all, there are still people who believe that the sun goes around the earth and who have strong arguments for this view. There are even people who believe that all species were individually created rather than arising by evolution!

is all against any general and continuous change of the environment alone that could account for the sort of change in life that has occurred. In particular instances of environmental change, evolving organisms do not respond in a uniform way, as if the environment were causing changes in structure. Moreover we now have a great deal of experience as to how heritable changes in organisms actually do appear, and there is the strongest sort of evidence that they do not appear as direct responses in kind to environmental influences.

On the other hand, it has already been shown that inherent tendencies for organisms to evolve in fixed directions regardless of the environment cannot be the whole story either, and more evidence against this view will appear as the enquiry continues. The most promising clue seems to lie in the interrelationship between organism and environment, and not in one or the other of these alone. This functional interrelationship is adaptation, one of the major features of evolution and of life. The environmental presence of dissolved oxygen in water and the internal physiological system of fishes based on intake of oxygen are interrelated by the adaptation of gills (and a whole complex of associated structures), which remove oxygen from the water and supply it to the internal system of the fish. The marvelous mechanism of the human hand is an adaptation to perform functions eminently useful in the human environment. The fangs of a wolf are an adaptation to environmental requirements for defense and offense. Examples of such relationships are so extremely widespread and so well known to everyone that it is needless to multiply them here.

The key to the adaptive nature of any characteristic of an organism is its usefulness under the conditions in which the organism lives. Just here it does not particularly matter how this relationship arises, although of course that is one of the main objects of our enquiry. Whatever we may later decide to be its cause, the fact is that adaptation, by the criterion of usefulness, certainly does occur. Every organism has innumerable adaptive characteristics. This is one of the most universal features of living things. The present point is to decide whether adaptation is the orienting factor in evolution. If so, continuing trends in evolution should be adaptive in character. The changes involved should be useful to the organisms. This usefulness may be in better fitness for the way of life already followed by the organisms, or may enable them to cope with imposed changes in that way of life, or may permit (or accompany) shift to some other way of life.

Examples make these possibilities more concrete. Ability to run faster, a trend in herbivore evolution—although as usual, not universal or constant—is evidently useful, that is, adaptive, in the way of life of animals dependent mainly on speed for safety. As their prey thus becomes swifter, a change in the way of life of carnivores is imposed on them: it becomes advantageous, adaptive, for them also to run faster or to develop attack from ambush or other ways of coping with the increased speed of the herbivores. These are trends observable in the orientation of carnivore evolution, in various lines. Among herbivores, again, as competition for food becomes severe among the browsers, living on succulent or leafy herbage, it becomes advantageous for some of them to shift to a different way of life, eating some other sort of food such as grass. This adaptive trend to a new sort of adaptation is prominent in one phase of horse evolution, when high-crowned grazing teeth arose. (It is typical of such shifts in adaptive type that the phase was short, the shift relatively rapid.) Some carnivores showed analogous shifts, with adaptive trends toward new ways of life as regards food habits; an extreme example is the rise of a strictly vegetarian "carnivore," the giant panda.

If, on the other hand, adaptation is not the only or at least the dominant orienting factor in evolution, then there should be recorded clear-cut and probably numerous examples of trends that are not adaptive in nature. We should expect to find long-continued changes of no apparent use to the organisms. The qualification "long-continued" is necessary here because there certainly are shorter, random, nonoriented phases in evolution, which constitute a distinct problem for later discussion. We should expect, especially if we happen to be finalists, that eventually useful characteristics may appear and become trends before they are useful. We should also expect some trends to persist without reference to changes in habits of the animals exhibiting them. And perhaps most crucial of all, there should be trends, possibly useful in earlier stages, which persist until they are of no further use or become finally downright harmful. These may even be expected to cause extinction in many cases.

No one doubts that there are adaptive trends. These surely exist and are numerous. The question is as to the existence or prevalence of nonadaptive trends. Every type of nonadaptive trend mentioned above has been claimed as existent by students who do not think that evolution is oriented by adaptation, and supposed examples of each have been given by them. It is notoriously impossible to prove that a rule has *no* exceptions, that, for instance, a nonadaptive trend *never* occurred, for we do not know and no one could examine the intricacies of all evolutionary trends whatsoever. It is also impossible in reasonable compass to examine each one of the many supposed examples of nonadaptive trends. It is, however, possible to study outstanding and crucial examples of the claimed sorts of nonadaptive trends and to base on these results a reasonable judgment as to the probability of nonadaptive orientation in evolution.[8]

8. These and other supposed examples may be found in abundance in the writings of vitalist evolutionists and philosophers, in the attacks on Darwinism and Neo-Darwinism common a generation ago and still continuing in some circles although largely abated, and in general in espousals of "autogenetic" theories of evolution, which hold that organisms are

Among supposed examples of nonadaptive trends, the sabertooth (or saber-toothed "tigers," but they were no more tigers than they were lions, lynxes, or tabbies), the Irish elk, the coiled oysters or gryphaea, the titanotheres, and the horses are typical, perhaps most often cited, and well cover the variety of such supposed phenomena. Even these must be treated briefly here, as each case is so complex as to merit book-length discussion.

The sabertooth is one of the most famous of animals just because it is often innocently supposed to be an indisputable example of an inadaptive trend. The poor sabertooth has come to figure as a horrible example, a pathetic case history of evolution gone wrong.[9] Its supposed evidence is thus characteristically summarized in a book on (human) personality: "The long canine tooth of the saber-toothed tiger grew more and more into an impossible occlusion. Finally, it was so long that the tiger could not bite effectively, and the animal became extinct."[10] Now, like so many things that everyone seems to know, this is not true. Sabertooths appear in the record in the early Oligocene, more or less 35,000,000 years

---

self- or auto-evolving and that the orienting factors are inherent in the organism in some unexplained way. An outstanding older example of this school is L. S. Berg's *Nomogenesis; or Evolution Determined by Law* (trans. from the Russian by J. N. Rostovstow [London, Constable, 1926]). One of the most reserved and reasonable and yet, as I believe, erroneous full-length statements of the finalistic, autogenetic point of view is L. Cuénot and A. Tétry, *L'Évolution biologique* (Paris, Masson, 1951). That final publication by Cuénot, one of the great biologists of his generation, is less confident of finalism than most of his earlier work.

9. For example, the previously cited book by Schindewolf includes miscomprehension of what is really known of sabertooth history.

10. G. Murphy, *Personality. A Biosocial Approach to Origins and Structure* (New York and London, Harper, 1947). Dr. Murphy is hardly to be blamed for accepting what was long a cliché among the less critical of paleontologists.

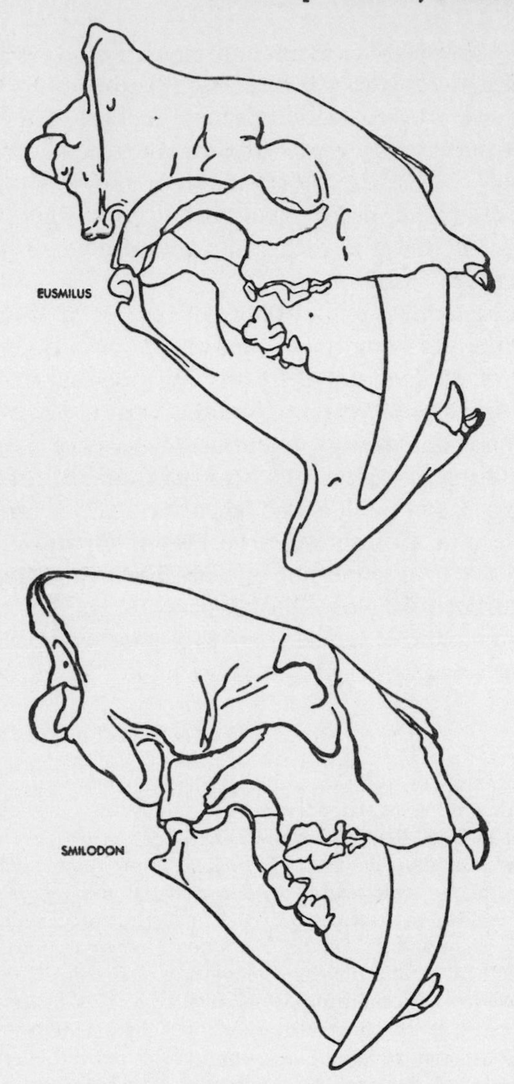

Fig. 26. Skulls of saber-toothed "tigers." *Eusmilus,* early Oligocene, is one of the earliest known forms and *Smilodon,* Pleistocene, one of the last. *Eusmilus* is smaller, but the two skulls have been drawn the same size to permit easier comparison.

ago, and they became extinct only yesterday, around the end of the Pleistocene, 20,000 or 30,000 years ago. The fact is that during this long span the sabertooths did not show a trend toward increase in relative size of the canine tooth. It happens that the earliest sabertooth known (*Eusmilus*) had relatively one of the largest canines known in the group. (See fig. 26.) Throughout their history the size of sabertooth canines varied considerably from one group to another but varied about a fairly constant average size, which is exactly what would be expected if the size were *adaptive* at all times and there were no secular trend in adaptive advantage but only local and temporary differences in its details. The biting mechanism in the last sabertooths was still perfectly effective, no less and probably no more so than in the Oligocene.[11] To characterize as finally ineffective a mechanism that persisted without essential change in a group abundant and obviously highly successful for some 35,000,000 years seems quaintly illogical! In short, the "inadaptive trend" of the sabertooth is a mere fairy tale, or more fairly, it was an error based on too facile conclusion from imperfect information and it has since been perpetuated as a scientific legend.

The case of the Irish elk (*Megaloceros*) is more seriously deserving of attention in this connection. This animal represented the extreme development of a group, offshoot of the common European stags (not really elks), in which the size and complexity of the antlers increased until finally they were fantastic. It seems that yearly replacement of these masses of bone must have been a severe physiological burden and that the strain of carrying such a weight on the head must also have been severe. Although there is a total lack of

11. On this and the basic features of cat evolution in general, see W. D. Matthew's classic paper, "The Phylogeny of the Felidae" (*Bull. Amer. Mus. Nat. Hist., 28* [1910], 289–316). There are numerous later studies of cat evolution which of course modify and add details, but they do not invalidate Matthew's broad thesis and no later over-all review of the group has been written.

decisive evidence that this development did, in fact, cause extinction of the group, it does seem possible that it was a factor in that extinction. There is nevertheless no real reason at all to think that the oversized antlers were inadaptive *when they arose;* it is conclusively evident that the big sabertooth canines were not inadaptive when they arose. The biggest antlered stags did not flourish as long as the sabertooths, only for a few thousand years, but they did flourish for a while. If the antlers were a factor in extinction, this was very likely because—as eventually, at long last, with the sabertooth canines—changing conditions made them inadaptive after they reached maximum size. In that case the trend up to that size was not inadaptive.

There is, however, an additional or alternative factor in the antler development of this group, that of relative growth. All animals have distinctive growth patterns determined in the main by heredity. In fact, what animals, and plants, too, inherit from their parents are not developed characteristics, such as long noses, blue eyes, or pink flowers, but determiners of a developmental process in the course of which such characteristics eventuate, more or less modified by the conditions under which development occurs. As growth goes on, all parts of the organism do not increase proportionately. Everyone has noticed that human babies have relatively shorter legs but bigger heads than adults. Some parts grow faster than the body as a whole, some parts at about the same rate and some more slowly. Some may even decrease in absolute size as the body grows: in later stages of human growth the thymus gland normally does this. Stags have a growth pattern in which as the body becomes bigger the antlers become even bigger; they increase not only in absolute size but also in size relative to the body of the animal. In such a group, if there were an evolutionary trend toward larger adult body size, there would also, automatically, be a trend toward relatively larger antlers. The Irish elk had the largest body of any of the deer in this group, and its antlers are just the size to be expected if the inherited relative growth pattern remained the same.[12] The real trend, then, was for increase in total body size. Antler size just tagged along, or raced ahead, in accordance with a growth pattern evidently adaptive in origin and adaptive in living stags. The body size increase in these great extinct stags was probably adaptive throughout, whether or not it carried antler size to an inadaptive degree, which remains a moot point.

It is the most reasonable view, at least, that increase in gross size, which is among the most widespread of trends in animal evolution, is oriented by adaptation. It is commonly advantageous for an animal to be slightly larger than others of its kind: the larger animal will be a little faster, a little stronger; in accordance with relative growth, it may have and be able to handle more effective weapons for offense or defense; it will, on an average, have more cells in the body, with increased opportunity for complication or differentiation of tissues and organs; under some conditions, such as cold climate for warm-blooded animals, it will have improved thermal efficiency.[13] The advantage need not be very great, and any considerable difference in size may be disadvantageous because, for instance, of disharmony in development, undue food needs, and disparity with other members of the species. We find, in fact, that the trend is usually slow and gradual, not in sudden steps. It also occurs in just those situations where we would expect it to be advantageous. On the other hand, there are conditions in which increase in size

12. Although the subject has been greatly elaborated and many more of its implications explored in later work, the best general treatment of relative growth, including that of antlers, continues to be J. S. Huxley's *Problems of Relative Growth* (London, Methuen, 1932).

13. The adaptive aspects of gross body size have been especially studied by Rensch, experimentally and otherwise, and some of his results are reviewed in his previously cited book.

is disadvantageous: in most burrowing animals or those depending for safety on obscurity and hiding in small places; in one-celled animals, above certain low limits, because there is no special means of circulation of metabolic supplies and products; in other animals for which (as there is finally for every animal) there is a limit of size beyond which mechanics and physiology cease to be efficient; in many types of flying animals (notably among insects) where the difficulty of supporting weight soon overbalances any advantage of greater size; in animals on islands or elsewhere where food supply is rigidly limited and greater use by each individual would reduce the size of the breeding population unduly. It is in just such groups that the "law" of size increase is suspended or repealed and that we find persistently small organisms, or reversal of trend from larger to smaller—as when elephants become stranded on islands. The rule of size increase in evolution operates in general as if it were thoroughly adaptive. If in special cases we cannot know for certain whether it is adaptive or not, we may properly assume that it is, unless clear evidence to the contrary exists. No such contrary evidence does exist in the case of the Irish stag or any other animal known to me. Throughout the world of life the adjustment of size to the effective following of a way of life is remarkably precise.[14]

The combination of adaptive size trends with relative growth produces simultaneous trends which may also be adaptive, or may be nonadaptive (of no particular advantage or disadvantage), or inadaptive (disadvantageous), as long as the disadvantage does not overbalance the advantage of being of the best gross size. This seems to account for

14. The argument, to this point, does not depend on whether the organisms have their way of life because they are the right size or are the right size because they have this way of life. I think neither statement is quite correct, but in any case the result is size adaptation.

many of the cases of apparently meaningless or disadvantageous features in oriented evolution. There are other ways, too, in which inadaptive change of a particular character may be merely a secondary result of a trend that is really adaptive.[15]

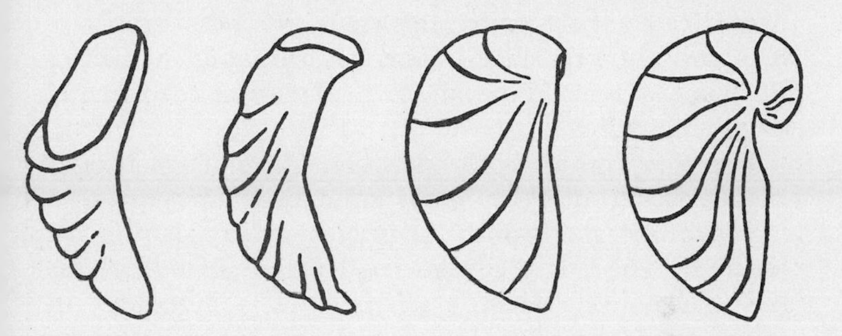

Fig. 27. Evolution of a coiled oyster, *Gryphaea,* right, from an oyster of more usual type, left. (Redrawn with some modification after Trueman.)

The example of the evolution of the coiled oysters has analogies with that of the sabertooths and of the Irish elk and need not now detain us long. At various times during the Mesozoic several different lineages of oysters recurrently developed a trend toward curling of one of the two shells into a plane spiral. (See fig. 27.) The highly coiled types are collectively called *Gryphaea.* In each case the trend is short

15. For discusison of some of these other rather more abstruse processes see, for instance, Rensch's book again or my *The Major Features of Evolution,* already cited. There is also some discussion of this aspect of trends in Huxley's important work, *Evolution. The Modern Synthesis,* also previously cited, which in every chapter has much that is pertinent to our present enquiry.

in geological time. It went on rapidly until the shells had one-and-a-half turns, more or less, in the oldest shells, and then each line became extinct. In extreme cases the coiled shell grew around until it pressed on the other shell and they could not be separated. When this happened the animal inside must have died, because it needs to separate the shells slightly in order to take in food and water and to expel waste products. According to one opinion, this trend was not adaptive or, if adaptive in early stages, was carried so far as to cause extinction because all the animals sealed themselves up in their shells and so died.

The trend probably was adaptive. A coiled shell is advantageous for an oyster in soft mud, and that seems to be the environment in which the gryphaeoid trends primarily occurred. There is really no good reason to believe that the change was carried to an inadaptive degree by the trend. Death by self-immurement came only to extreme variant individuals in their old age. As far as evolution is concerned, among nonsocial animals what happens to nonbreeders or to old animals that have finished their share of breeding makes no difference at all. (It is one of the radical differences in evolutionary processes caused by the development of social structures that in these the nonbreeders and those past breeding age may have a decided influence on evolution of the group.) It is also probable that a relative growth factor is involved and that coiling became stronger and had an earlier onset as the final size of the animals increased. In that case the last evolutionary stage represented a compromise between the best size, perhaps not yet reached, and the best degree of coiling, perhaps already somewhat exceeded. In any case, commitment to so special a type of structure or compromise between two concurrent trends may well have caused extinction as conditions changed and competitors developed; but this does not warrant the conclusion that the trends, while they continued, were anything but strictly adaptive.[16]

Supposed instances of nonadaptive trends in evolution may also concern the beginning rather than the end of such trends as in the preceding examples. The suggestion that characteristics later of great usefulness to their possessors may arise without any adaptive value and increase in a trend before this value develops is particularly related to the finalist thesis. The titanotheres provide a typical and often-cited example of this supposed phenomenon. These are an extinct group of Eocene and Oligocene hoofed herbivores belonging to the mammalian order Perissodactyla along with the horses, tapirs, and rhinoceroses. Their size increase was relatively rapid and was spectacular. In the early Eocene they were small, comparable to early horses in size, but in the Oligocene they were elephantine in bulk. In America they became extinct after the early Oligocene, but a few forms survived into the middle Oligocene in Asia. Their whole history was run in the course of perhaps 20,000,000 years, while the horse family, which evolved less radically in size but more in most other features, has a history of more than 50,000,000 years and is still going. The latest titanotheres had, in most cases, large protuberances of the skull bones on the front of the face which look like blunt horns and are commonly called horns, although the chances are that they were covered with calloused skin rather than true horn. The earliest titanotheres had no such protuberances. (See fig. 28.)

It seems quite apparent that these structures were adaptive. They must have been effective battering rams and useful as such to their possessors, whether fighting among themselves or fending off enemies. A skeleton in the Amer-

16. A more extended discussion of this example, with references to other literature, is given in T. S. Westoll, "Some Aspects of Growth Studies in Fossils." (*Proc. Roy. Soc., B, 137* [1950], 490–509).

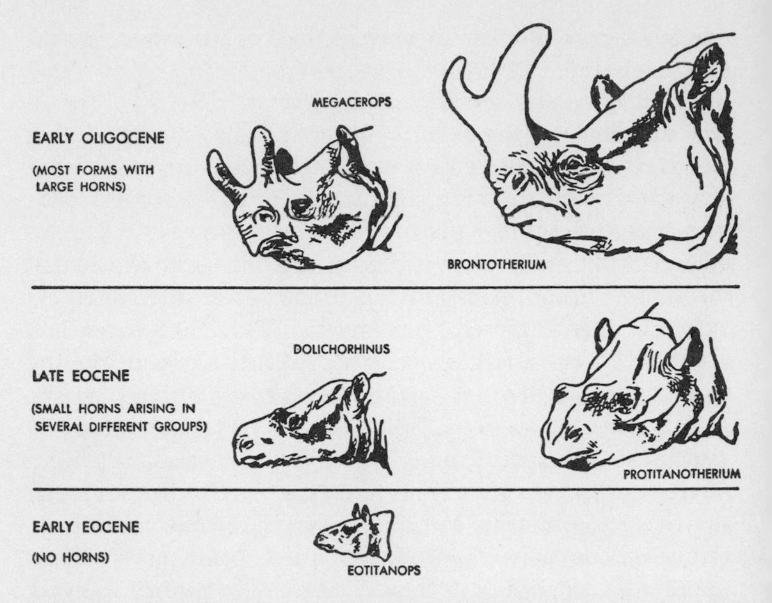

Fig. 28. Evolution of titanothere horns. Only a few of many different lines are shown. (Data from Osborn.)

ican Museum has a broken and later healed rib that could well have resulted from attack by the horns of a fellow titanothere. When horns first appeared, in the later Eocene forms, they arose independently in several different lines and arose gradually, at first as mere thickenings of the skull bones in this region. From this stage they developed into large, fully effective weapons by a rather rapid trend in each line.[17]

Argument that this trend was not oriented by adaptation depends on the assertion that the incipient stages of bone thickening were of no advantage or use to the animals possessing them: the structure starts first, and its usefulness

17. H. F. Osborn, *The Titanotheres of Ancient Wyoming, Dakota, and Nebraska* (U. S. Geological Survey, Monograph 55 [1929]).

comes later.[18] There are, nevertheless, at least two alternative explanations both of which would make the trend adaptive throughout and either of which or both together are more likely than nonadaptive orientation of the trend. The first consists simply of the counterassertion that the incipient stages were useful. It is likely that the animals had the habit of butting each other and their enemies, even at this stage; they had already become rather stocky, lumbering creatures with stout heads, and they had no other evident means of fighting. Thickening of the bones in the butting region would then be of advantage. The claim that the thickening was not enough to be useful at all or to be effective in orienting further change is not valid in the light of present knowledge. Any thickening would be of some advantage, however slight this might be, and many studies have now shown that in populations of medium to great abundance any appreciable advantage, even though exceedingly slight, may be surprisingly effective in producing further change of the same sort in subsequent generations.

There is also evidence that relative growth affected the evolution of horns in titanotheres.[19] The group seems to show a relative growth pattern in which below a certain body size horns do not occur at all, while above this critical

18. It is probable that this may occasionally happen in the case of some characters developed all at once by mutation. This is a sort of preadaptation and is one of the random, not one of the oriented aspects of evolution. Continued development, a trend, cannot be random throughout, although it may start at random. The question here is not whether a character ever appears prior to its usefulness but whether or not trends are oriented by adaptation. Still another problem concerns the transformation of a character developed with one use so that it comes to serve another.

19. See A. H. Hersh, "Evolutionary Relative Growth in the Titanotheres" (*Amer. Naturalist, 68* [1934], 537–561). Later studies of relative growth suggest modification of some of the methods used in this work, but its essential conclusions seem to stand.

size as growth increases the horns increase still more rapidly. Now if this growth pattern were already established before the later Eocene, the smaller animals of those earlier times would not have horns, but horns would appear in any and all lines in which adaptive increase in size passed the critical point. Thereafter development both of horns and of body size would be accelerated because both are advantageous, adaptive trends and the two go together. Thus the whole trend is adaptive from beginning to end. This begs the question, for the present, at least, as to how such a growth pattern arose in the first place, but there is certainly no reason to think it arose by a nonadaptive trend.

The example of the evolution of the horse family has been advanced as typical of a sort of over-all, nonadaptive control of trends. It has been claimed that in the horse family the major trends continue unvaryingly and unchecked regardless of changes in habit or in environmental factors. We have already seen enough of the facts of evolution in this family to suspect that this is not true. Brief further consideration fully confirms the suspicion. There are no really uniform and general trends in the evolution of this family; all trends are to some extent peculiar to one line or another, or effective only at one time or another. The most nearly general trends, such as progression in brain structure, would be useful and adaptive to all horses whatever their precise habits or environments. (In later stages, at least, this also involves correlation with size and a probable relative growth factor.) Increase in size was absent during the Eocene and in some lines at later times, and it was reversed in other lines. Where it does occur is apparently where it was adaptive; its very irregularity negates any over-all trend not adaptive in nature. In other characters, changes in habits or environmental conditions, to the extent that we can infer these from the record, do definitely involve also changes in the direction of trends. Indeed the whole history of the family seems outstandingly clear in evidence that its trends cannot possibly be explained by control by any factor except adaptation.

All these examples, and many more of the same sort cited by opponents of this view, seem on even this superficial analysis rather to support than to oppose the view that adaptation is the orienting factor in evolutionary trends. At the least, all such trends could possibly have this basis. Still more examples could be given to show that trends stop or change direction when they cease to be advantageous, in other words, when the adaptive control dictates. This happens, indeed, as just mentioned for the horse family, in the very examples often put forward to prove that it does not occur!

Adaptation has a known mechanism: natural selection acting on the genetics of populations. More as to the operation of this mechanism will be said when more has been learned about adaptation and other evolutionary phenomena. It is not quite completely understood as yet, but its reality is established and its adequacy is highly probable. In seeking the orienting factor in evolution we have seen that in some cases this must, by all reasonable inferences, be adaptation and in all, even the most doubtful, it could be adaptation. Thus we have a choice between a concrete factor with a known mechanism and the vagueness of inherent tendencies, vital urges, or cosmic goals, without known mechanism.

With no pretense at having plumbed the whole mystery or excluded all other possibilities, it is concluded that the major (if not the only) nonrandom, orienting factor in the process of evolution is reasonably identified as adaptation. Some examples have been given and some implications suggested. The enquiry is to be pressed forward into other aspects of the complex phenomena of life.

XII. THE OPPORTUNISM OF EVOLUTION

Over and over again in the study of the history of life it appears that what can happen does happen. There is little suggestion that what occurs *must* occur, that it was fated or that it follows some fixed plan, except simply as the expansion of life follows the opportunities that are presented. In this sense, an outstanding characteristic of evolution is its opportunism. "Opportunism" is, to be sure, a somewhat dangerous word to use in this discussion. It may carry a suggestion of conscious action or of prescience in exploitation of the potentialities of a situation. Language developed (it, too, is a phenomenon of the evolution of life) in a strictly human setting, for communication between the members of mankind. Its words too often carry undertones appropriate to the human scene and misleading in discussion of the grander scene in which men have so late, so brief, and yet so important a part. The present discussion is a communication from one human being to others, so it employs the words already developed for this purpose. It tries to avoid the amiable foible of scientists who are so prone to coin new words for phenomena which should not be viewed in terms of human motives and actions. But when a word such as opportunism is used, the reader should not read into it any personal meaning or anthropomorphic implication. No conscious seizing of opportunities is here meant, nor even an unconscious sensing of an outcome. The word is only a convenient label for these tendencies in evolution: that what can happen usually does happen; changes occur as they may and not as would be hypothetically best; and the course of evolution follows opportunity rather than plan.

What can happen is always limited and often quite strictly limited. Boundless opportunity for evolution has never existed. This has already become apparent in the study of orientation and trends in evolution. Possible ways of life are always restricted in two ways: the environment must offer the opportunity and a group of organisms must have the possibility of seizing this opportunity. On both sides these possibilities have changed greatly in the course of evolution and have, on the whole, become more abundant and more varied. Life, opportunistically, has become more abundant and more varied in accordance with the expansion of these possibilities. Life itself has made the greatest difference in expanding the possibilities of the environment, for each change by one group has opened new opportunities for others: the rise of land plants created the earlier nonexistent opportunity for the development of land animals; the rise of vertebrates provided, in the blood streams of these animals, a new and favorable environment for protozoan life. Purely physical expansions of environmental opportunities have often occurred, too, although many of these have turned out to be temporary. The uprearing of mountain heights provides the opportunity for differentiation of plants and animals adapted to more rarified atmosphere, more intense solar radiation, more constant cold, and other alpine conditions. The spread and retreat of glaciers or the permutations of climates between warm and cool, moist and arid, seasonal and equable, also provide continually changing opportunities.

On the side of the organisms, each change in them also changes their possibilities for future development, usually limiting them on one side and expanding them on another. Mere geographic position may do this: opportunity for expansion into the sea does not exist for land forms that do not reach the shore; opportunity for alpine expansion may arise for groups that spread into the foothills. Changing structure has still more pervasive and profound effects on limitation and expansion of possibilities. Agglomeration of cells, the origin of multicellular or metazoan structure, put a new lower limit on size and excluded many ways of life followed by the protozoans, but it greatly increased the upper limit on size and it began to open up a new world of opportunities involved in the increasing possibility of tissue and organ differentiation. Increasing development of grasping feet in primates excluded for them any immediate opportunity to match the fleetness of the deer, but it led to opportunities whose seizing eventuated in the finger on a trigger, against which the deer's fleetness is futile.

Within the groups of organisms there are other universally potent controls over what changes can happen in their evolution. They must have the materials for such changes, not only in the sense of established structural norms but also and more particularly in the sense of variations from those norms, of the potentiality for changes in them in order to adapt them to meet new opportunities or necessities. One way in which organisms meet this need is by flexibility in their existing growth patterns and functional systems. Twin seeds from a tree may be planted, one at timberline and another on a watered plain; if the growth pattern common to the two is sufficiently flexible and adaptable, the first will produce a matted shrub well suited to alpine life and the second will produce a tall and slender tree ideally adapted to its more genial situation. An animal whose dental and digestive systems are omnivorous in type may change from one type of food to another, as opportunity presents, without requiring or awaiting an evolutionary change in those systems.

Conversely, a seed from a tree whose growth is rigidly patterned to life on a fertile lowland will not grow if planted on a mountain top, and an animal with teeth and digestive tract adapted strictly for eating meat alone cannot, unless radical structural change intervenes in the evolution of the group, meet a need or exploit an opportunity to live on fruits. Every group of organisms has its established possibilities in these respects, sometimes a very wide range and sometimes very narrow, but always with limits. A bear may be nourished by ants, fish, chipmunks, raspberries, corn, and a great variety of other foods, but still it cannot survive on leaves or grass. A weasel must have fresh blood and flesh or it soon dies. A cow must have green vegetable food, but almost any sort of leaves or grass permits survival. A koala requires eucalyptus leaves and no others will do.

Of importance still more profound are possibilities and limitations not, like these, inherent in the individual or in the norm of the group but permitting or prohibiting the rise of new kinds of individuals and of new norms. Such developments are the real crux of evolution. (We do not suppose that bears have evolved when they stop fishing and range inland to eat berries, although the ability to do this is an essential factor in bear evolution.) In a larger sense, the development of opportunities in evolution demands changes in the heredity of the given group. Such changes involve three processes: the determinants of heredity, mainly the genes and chromosomes, already present in the group may change in proportions within the population, some perhaps being eliminated, others once rare becoming common; these determinants may be recombined in different ways and so reinforce or modify each other that their result, the growing and adult organism, comes to be of a type not previously present in the population; or new sorts of determinants—mutations—may appear within the population. All three processes are usually going on at the same time in any continuing, reproducing group of organisms and all three are involved in the usual long-range events of evolution.

Opportunities for evolutionary change are thus ultimately both created and limited by what can result from determinants already present in a population of organisms and from mutations among those determinants, which, in fact, are themselves the results of earlier mutations. Opportunities for flying insect eaters existed for millions of years before there were any pterodactyls, birds, or bats to exploit them. The opportunity could not be seized until there were organisms that had certain mutations and populations that had integrated these into genetic systems necessary for development of wings and of the many other related features of this way of life. The immediately required mutations and genetic systems had to follow after innumerable others for development of limbs that could become wings, of jaws that could be adapted to seizing insects, and so on through an almost incredibly complex sequence of changes.

Among large numbers of organisms and over long periods of time such as are involved in the history of life, mutations are exceedingly abundant, but they are also exceedingly erratic. In experimentation, some ways of speeding up their appearance and of slowing it down have been discovered, but no general principle as to the orientation of their effects has yet been found. We do know one negative fact: the results of mutations do not tend to correspond at all closely with the needs or opportunities of the mutating organisms. It is a rather astonishing observation that the supply of this basic material for evolution seems to have no particular relationship to the demand. This accounts for much of the opportunism in evolution, and the nature of that opportunism in turn attests the random nature of mutation.

Evolution works on the materials at hand: the groups of organisms as they exist at any given time and the mutations that happen to arise in them. The materials are the results of earlier adaptations plus random additions and the orienting factor in change is adaptation to new opportunities. If this view of evolution is correct, then we must expect to find similar opportunities exploited in different ways. The problems involved in performing certain functions should have multiple solutions. As environmental opportunities arise, it should be seen that their exploitation is occasionally delayed or even, rarely, missed altogether. Furthermore, when the opportunities are seized, that may be by a variety of different groups all of which will attain similar adaptations but may do so along different paths and by different means. These expectations are, in fact, abundantly fulfilled in the history of life.

Take, for instance, the expectation of multiple solutions of adaptational problems. These exist for almost any such problem that you can name and they account for much of the swarming diversity of life. The antelopes probably provide an example as good as any among the great number that could be adduced. An essential feature of their adaptation or requirement of their way of life is the development of horns, their principal means of offense in struggles within the groups and of defense against attack from other animals. All male antelopes, at least, have horns, but they are not exactly alike in any two species. Even among the antelopes of a single region today, with diversity far less than for the group as a whole in its far-flung history, the differences are quite amazing, as may be seen in figure 29.

Now there must be some one type of horn that would be the most effective possible for antelopes, with some minor variation in proportions or shape in accordance with the sizes or detailed habits of the animals.[1] Obviously not all of

1. There are cases, particularly among birds but also in other groups, where it is advantageous for conspicuous markings or structures to be different among otherwise similar animals, as means of quick recognition by their fellows in the same species. If this were an important factor among antelopes, the horns should be distinctly different among the forms most similar in other respects and so most likely to be mistaken

Fig. 29. Heads of the principal types of antelopes in the Belgian Congo. For discussion, see text. Direct copy from H. Schouteden, "De Zoogdieren van Belgisch-Congo en van Ruanda-Urundi," *An. Mus. Belg. Congo,* C, ser. 2, 3 (1947), fasc. 1–3.

these antelopes have the "best" type of horns, and probably none of them has. Why, for instance, with their otherwise rather close similarity, should the horns of the reedbuck (14 in the figure) curve forward and those of the roan antelope (18) curve backward? Do not the impala (11) and kob (15) horns, with their double curve, seem to achieve the same functional placing and direction as the reedbuck horns (12 and 14) with a single curve, but to do so in a way mechanically weaker? Even though the animals themselves are small, are not the duiker horns (9 and 10) too small to be really effective, and are not the tremendous kudu horns (23) unnecessarily unwieldy?

If evolution were really operating according to a fixed plan, surely these radical discrepancies would not arise. Nor would they arise if evolution were basically orthogenetic or if a rigid orientation were everywhere the rule. It looks as if there had been an orienting factor, indeed—all these animals do have horns that serve them well enough even if not perfectly—but as if what it had to orient was not uniform or completely responsive. As horns developed in the different lines different sorts of mutations affected them. As long as the mutations increased the development of a functional horn, however various in other respects, they served the adaptive end and were retained and promoted in the evolution of each line. The mutations that would have been mechanically the best, that an engineer would have chosen, simply did not happen to occur in all, or perhaps in any, of these lines. There are two aspects of opportunism: to seize

---

one for the other. Such is not the case; species most similar in other respects are also most similar in the horns. Note, for instance, the two species of reedbuck (*Redunca*), 12 and 14 in the figure, or of sable or roan antelope (*Hippotragus*), 18 and 19. Even if recognition were a factor in the evolution of these horns, the nature of their diversity would still be inexplicable as completely adaptive.

such diverse opportunities as occur, and when a single opportunity or need occurs, to meet it with what is available, even if this is not the best possible. The antelopes, and many other groups of animals, well illustrate this second sort of evolutionary opportunism.

Examples like this have, oddly enough, been used to argue that horns (or whatever feature it may be) are not adaptive and that their usefulness, if any, is sheer accident. It is claimed that adaptive control should mean the reaching of a single solution for each functional problem. This quite overlooks the established fact that adaptation can only orient changes that really do arise in animals and that these are guided but not evoked by adaptation. It is only under vitalist and finalist theories that one can suppose that the changes that arise are indeed just the ones needed for best adaptation —and examples such as that of the antelopes add their weight to the great bulk of evidence making those theories untenable. On the other hand, if under the claimed alternative there were no orientation whatever and horns developed quite by chance, then surely some would be merely fantastic and would not work at all. They all do function sufficiently well, some better and some worse but all well enough to serve. It really seems as if the explanation given here were the only one that is reasonable.

Another example is worthy of some discussion not only because it reinforces the conclusion suggested by antelope horns but also because it has an important bearing on some other important evolutionary principles. This is the nature and evolution of photoreceptors among animals. In a great many animals, invertebrate and vertebrate, it is clearly advantageous to have some means of receiving and consequently of responding to the stimulus of light. In some animals this may be of no particular value and in a few it may be downright disadvantageous, but it is a characteristic useful in a particularly wide variety of environments and of ways of life. Apparatuses for receiving light stimulus, photoreceptors, have correspondingly developed in a great number of widely different sorts of animals. Some photoreceptors merely give the information that light is present, or absent. Others give, sometimes with great accuracy, information as to the direction from which light is coming or as to the relative intensities of light from different directions. Still others form an image and give information about the shape of the particular object that is emitting or reflecting the light received. They may even add information as to how far away the object is, how fast it is moving, and what wave lengths of light (colors) it emits or reflects. Our own eyes do these things and are among the most versatile and effective ever evolved, although there are many animals with eyes that excel ours in some particular or another.[2]

The function of receiving light stimuli is highly special and definite in its requirements. It might be supposed that there was just one way to develop this function, or at least that there was one best way leading at last to an apparatus capable of receiving all the information that light can convey. In fact the variety of photoreceptors among animals is almost incredible. Given the properties of animal cells and tissues in general, it seems nearly impossible to think of a practical means of photoreception that has not appeared in one group or another, or in more than one. (For a very few examples, see figures 30–31.) And yet the theoretical best or the perfect eye has never been achieved. All fall decidedly short of the ability to gather all the informa-

2. One of the most useful discussions of photoreceptors in general is in L. Plate's *Allgemeine Zoologie und Abstammungslehre*, Teil II: "Die Sinnesorgane der Tiere" (Jena, G. Fischer, 1924). A truly fascinating account of the structure and history of the particular sort of image-forming eye that arose early in vertebrate evolution is given by G. L. Walls, *The Vertebrate Eye and Its Adaptive Radiation* (Bloomfield Hills, Michigan, Cranbrook Institute of Science, 1942).

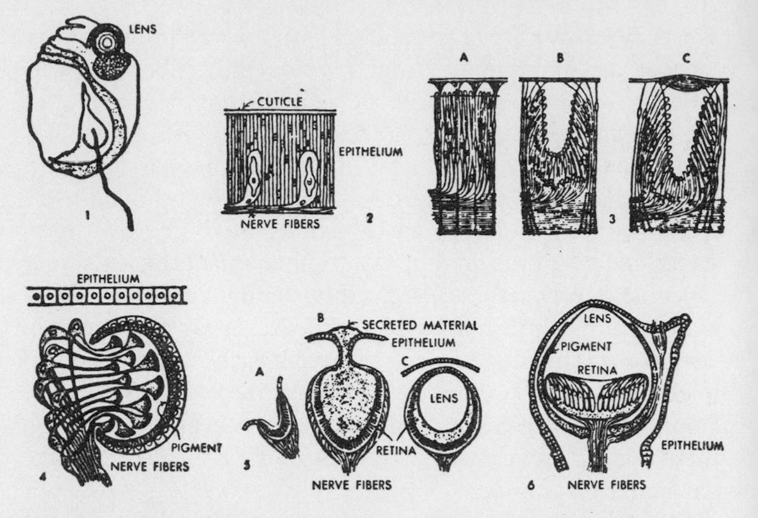

Fig. 30. Some light-receiving organs or photoreceptors. 1, a single-celled or noncellular protozoan *(Pouchetia)* with a light sensitive pigment-spot (stigma) and lenslike denser region. 2, section through skin of an earthworm with two of the scattered light-sensitive cells. 3, sections through three types of photoreceptors in starfish: A, sensitive cells grouped in a flat plate; B, cells infolded to form a basin; C, basin with rudimentary lens. 4, section through photoreceptor of a flatworm, with ends of sensitive cells pointing into a pigmented basin. 5, sections through three types of eyes in gastropods (snails, etc.): A, a simple basin; B, deep retinal basin filled with lenslike secretion; C, well-developed spherical eye with retina and lens. 6, section through the highly differentiated eye of a clam.

tion that light can convey. The best of them have had to compromise, sacrificing one sort of information for better reception of another.

In some protozoans the body, undivided into separate cells, is light-sensitive as a whole. In others a special photosensitive spot (stigma) is developed within the protoplasm. Some of these forms also have a lenslike mass of high re-

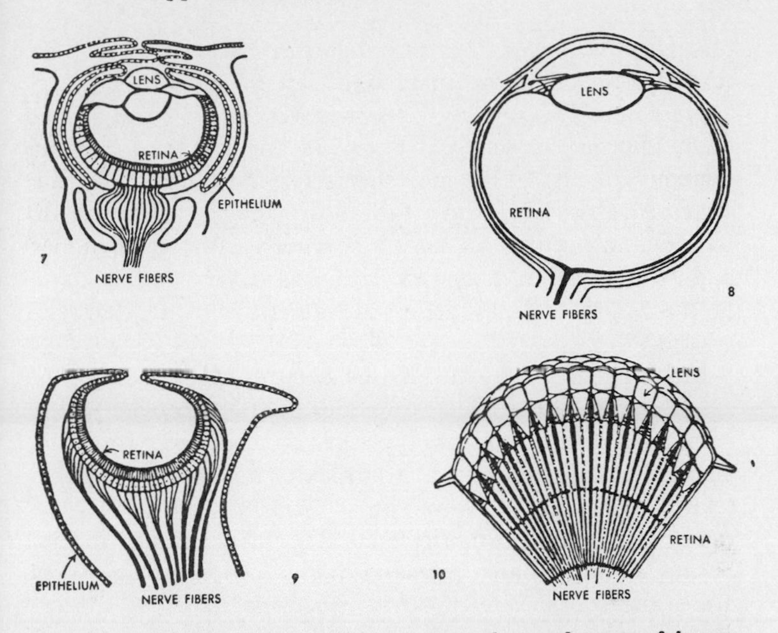

Fig. 31. Continuation of fig. 30, some advanced types of image-forming eyes. 7, section through the eye of a cuttlefish, functionally like the eyes of vertebrates but developed differently. 8, section through the eye of a vertebrate (man). 9, section through the pinhole eye of the chambered nautilus. 10, section and partial perspective of the multiple-tube (compound) eye of an insect (diagrammatic, only). (Figs. 30–31 have been redrawn from numerous sources, data mostly as compiled by Plate after Hesse and others.)

fractive index that concentrates light on the sensitive spot. These types seem to exhaust the possibilities without cell and organ differentiation. More numerous and complex types of photoreceptors appear among the multicelled animals. Special photosensory cells may be scattered over or in the body. These may be gathered in groups, which may vary greatly in size, number, shape, and position. More distinctly defined groups of these photosensory cells may form flat plates, grooves, basins, or bubblelike vesicles. They may be turned toward the incoming light, away from it, or at an angle to it. The sensory cells may be few or exceedingly many and may or may not be backed by a special layer of pigmented cells, which may themselves be of most various forms. All these types and combinations may be accompanied by a light-concentrating lens (again with a variety of shapes and structures), or may lack this. Moreover, similar parts of the apparatus may arise in different ways in different groups.

Image-forming eyes,[3] which gather information not only about incoming light but also about the objects from which it comes, are also of various types but are confined to animals of considerable anatomical complexity and are themselves relatively complex photoreceptors. Aside from the fact that image formation is complicated and requires a large number of highly differentiated parts to function properly, it requires also some complexity of other functional parts of the organism if the image so formed is to be of any particular use. If a protozoan were to receive an image by its sensory apparatus it is, so to speak, difficult to see what it could do about this; the protozoan lacks sufficiently complex systems to discriminate one sort of image from another or to modify its behavior accordingly, nor is it conceivable that such complexity could be developed within a single mass of protoplasm. Similarly among other animals, when photoreceptor evolution stopped short of image formation it seems likely, at least, in most cases that this was because it had gone about as far as was really useful (or adaptive) for the particular sort of animal, with its general structural level and way of life.

Image-forming eyes are by no means uniform, even as regards the general optical principles on which they oper-

3. Some students call these "iconogenetic," but the term is no shorter than "image-forming" and means exactly the same thing to fewer people.

ate. There are optically four ways in which images can be formed in natural light: lens, pinhole, multiple tubes, and concave reflector. The first three are all found in image-forming eyes of various animals, the lens, for instance, in you, the pinhole in the chambered nautilus, and the multiple tubes in flies. The reflector principle does not seem to have been used in any image-forming eye unless, just possibly, in some extinct type. It is, in fact, improbable that animal anatomy and tissues could have met minimum requirements for efficient use of the reflector principle.[4] The lens and pinhole principles are familiar enough to photographers and most others. They may operate simultaneously, as when a camera diaphragm is stopped down, and this combination is also common among image-forming eyes. (The iris is the diaphragm in the eye; its contraction stops down the opening and increases depth and sharpness of focus—in some animals with lens eyes going so far as essentially to take over the image-forming function from the lens.) Multiple tubes operate on the principle that since light travels in straight lines, only light from straight in front of each tube will traverse it and come out the other end. If each tube of a large bundle is pointed straight at a different part of the object, each will transmit light from that part only, and place it in a mosaic that is, at least roughly, an image of the object. The system would not be flexible or efficient for photography but it is employed, and highly developed, in the compound eyes of insects.

Lenses arose in evolution as light-gathering or -concentrating rather than as image-forming apparatuses. Since they can, in fact, create images as well as gather light (and, in-

4. Retina and pigment cup or disc would have to be reversed with respect to incoming light and interposed in its path without cutting it off, while an efficient internal reflector would have to develop. Such a development might not be quite impossible but it seems very nearly so.

cidentally, pinholes and multiple tubes cannot gather light), it is not surprising that they were most often involved in further evolution to image-forming eyes. In the evolution of compound eyes they were retained but did not change function: in the insect eye each tube has a lens which serves to concentrate and direct light, not to form an image. In molluscs and in the vertebrates, two groups with optically advanced image-forming eyes, the lens usually became the image former.

The remarkable similarity between the eyes of an octopus (which is a mollusc) and of a man has been cited as evidence of the existence of a uniform plan in evolution, of some over-all life force or final aim. Yet we have seen that these are only selected examples among a great array of photoreceptors which include almost every conceivable sort. And also, hidden in these eyes there is an evolutionary joker which is evidence, too: they work in almost exactly the same way, but they do not develop and did not evolve in the same way. In the vertebrate eye the retina or rather its sensory cells are aimed away from the light and in molluscs they are aimed toward it. The same remarkable difference is found in some non-image-forming eyes among invertebrates. Evolution has been opportunistic again: among its myriad solutions of the problem of photoreception are two in which the solution in principle is almost identical but in which, as usual, it did only the best possible from the materials in hand. Those materials happened to be different in the two cases so that actual identity could not be achieved.

The evolution of the eye is frequently cited in opposition to the sort of conclusions here drawn by two schools of evolutionists who, aside from this, may have little in common. Their contention is not entirely pertinent to the present chapter, but it is desirable to mention it briefly here while the example is before us. Both schools submit that the image-forming eye could not function until *after* it was complete.

From this statement the two draw radically different conclusions. One group, the finalists, concludes that therefore the eye evolved with reference to a *future* function, that the structure began first and use for it came only millions of years later when it was fully evolved and could begin to fulfill the purpose for which it was destined from the start. The other group unites, on this point, such odd bedfellows as the antievolutionists or special creationists and certain evolutionists who disbelieve in adaptive control of evolution and believe instead in catastrophically sudden changes of structure—what one of them, Schindewolf, calls "typostrophism." These agree that since the image-forming eye, so they say, could not function until complete, therefore it became complete all at once. It did not slowly change to its present state through the ages but arose as it now is in one bang, created so by God or by Mutation—or Typostrophism.

Inadequate as it is, the preceding discussion of photoreceptors sufficiently points up the fallacy of these contentions. If all photoreceptors except the image-forming eyes of advanced molluscs and of vertebrates had become extinct, there would be an excuse for bewilderment. In fact, representative stages at every gradually different level happen to have survived, from diffuse photosensitivity of the whole body through scattered photosensitive cells to cell plates, basins, basins and vesicles plus lenses, and so on to the fully developed image-forming eye with lens, iris, and its other complexities. These photoreceptors function splendidly at every level and do not wait to start working until the final stage is reached. They simply enlarge, refine, and to some extent change their functions as they become more complex.

To return to more direct discussion of the opportunistic nature of evolution, this is still more widely apparent in such phenomena as divergence and adaptive radiation or convergence and analogous change from different origins. Adaptive radiation represents the exploitation of possible ways of life by the groups of organisms that happen to be present at the same time and place where the opportunities occur and by means of recombinations and mutations in the characteristics that they happen already to have. In Part I of this book it has already been seen how major, and also minor, groups of animals start with some basic adaptive type, the descendants of which then tend to diverge and to take on a wide variety of habits. It should be emphasized that the basic type is itself specifically adapted to a particular way of life. It does not broadly include all the ways later followed by its descendants, nor does it simply embody the general features that these may all have in common. As Romer has so well remarked, you cannot make a living by being generalized.

The basic placental mammal, for instance, occupied a particular niche in the ecology of its time. It was an active, tiny, quadrupedal type of animal, with its obscurity as main defense, eating mainly small animal food such as worms or insects and their larvae, but probably also somewhat omnivorous and able to supplement its diet with some of the more nutritious plant foods, small fruits and the like. From this basis, evolution opportunistically developed extremely diverse types. Some mammals continued to follow the ancestral way of life but became more highly and rigidly adapted to it and parceled out its various specialties among different lines. These are the living insectivores. Some became aerial (bats) and some aquatic (whales, and others). Some became herbivorous (most hoofed mammals, most rodents, and some others) and some carnivorous (most carnivores and some others). Each category can be repeatedly subdivided: the diversity includes hundreds of distinct adaptive types. Thus among the many sorts of herbivores some are terrestrial, some subterranean, some arboreal, some aerial, some aquatic. Among terrestrial herbivores some browse, some graze, some eat fruits, and so on. Among terrestrial grazers, some are small, some large; some are fleet, some ponderous. It seems as if no possibility had been skipped.

The essentially opportunistic nature of this process is particularly evident when, as has repeatedly happened, the same sorts of opportunities have occurred for different groups of animals. An often-repeated example is that of the marsupials as compared with the placentals. It happens that of these two major subdivisions arising among early (Cretaceous) mammals, only the marsupials soon reached Australia, then as now an island.[5] Within Australia from this marsupial stock developed a radiation that came to include most of the important adaptive types that arose elsewhere by radiation of the placental mammals. There are insectivorelike marsupials (for example, the so-called marsupial anteater, *Myrmecobius*), molelike marsupials (the pouched "mole," *Notoryctes*), doglike marsupials (the Tasmanian "wolf," *Thylacinus*), squirrel-like marsupials (the phalangers, subfamily Phalangerinae, including even marsupial "flying squirrels," the flying "possums" or gliders of the Australians, *Acrobates* and *Petaurus*)—and so on, through many more types. The variety is not as great as among the placentals, but the opportunities offered by Australia were obviously not as exten-

5. Some geologists deny that Australia was an island when mammals spread to it, and the whole history is sometimes interpreted quite differently. There is, nevertheless, excellent and, I think, adequate reason for the views here briefly suggested. Other complications are introduced by the fact that sometime during the Tertiary two types of placental mammals did gain access to Australia and become abundant there: the Old World rats and the bats. In the Pleistocene came man and the dog (dingo). This is not the place to consider the whole very large problem of the history of the Australian fauna. The differences of opinion and the complication do not really matter here, anyway, because there is general agreement on the main point of analogous radiations among marsupials and placentals. The history is reviewed in my paper, "Historical Zoogeography of Australian Mammals (*Evolution, 15* [1961], 431–446) and book, *The Geography of Evolution* (Philadelphia, Chilton, 1965).

sive or varied as those offered by all the other continents put together.

South America, also an island during most of the Tertiary, provides an even more extensive example, complicated by the fact that its original mammalian stocks were more varied than those of Australia and included both marsupials and placentals. Lengthy discussion is not necessary, but a few of the products of this radiation may be compared in figure 32 and in tabular form with those of North American radiation, which was also connected and involved with Eurasia and Africa at various times:

SOUTH AMERICA		NORTH AMERICA	
General stock	*Special types*	*Special types*	*General stock*
Marsupialia	Caenolestines	Shrews	Insectivora
	Polydolopids	Squirrels or mice	Rodentia
	Borhyaenines	Dogs	Carnivora
	Thylacosmilus	Sabertooths	
Pyrotheria	Pyrotheres	Mastodonts	Proboscidea
Litopterna	Macraucheniines	Camels	Artiodactyla
	Proterotheres	Horses	Perissodactyla
Notoungulata	Toxodonts	Rhinoceroses	
	Homalodotheres	Chalicotheres	

That such phenomena represent opportunism and not plan or purpose in evolution is attested by the fact that such independent radiations produce animals similar in ways of life but not the same in structure, indeed sometimes remarkably different. The kangaroos of Australia (along with a few extinct types such as *Diprotodon*) are analogous to the hoofed herbivores of the rest of the world and have exploited some of the same opportunities as far as these existed in Australia—but how obviously different they are! The Tasmanian "wolf" or thylacine, at the other extreme, does look almost exactly like a dog of sorts at first sight, and yet

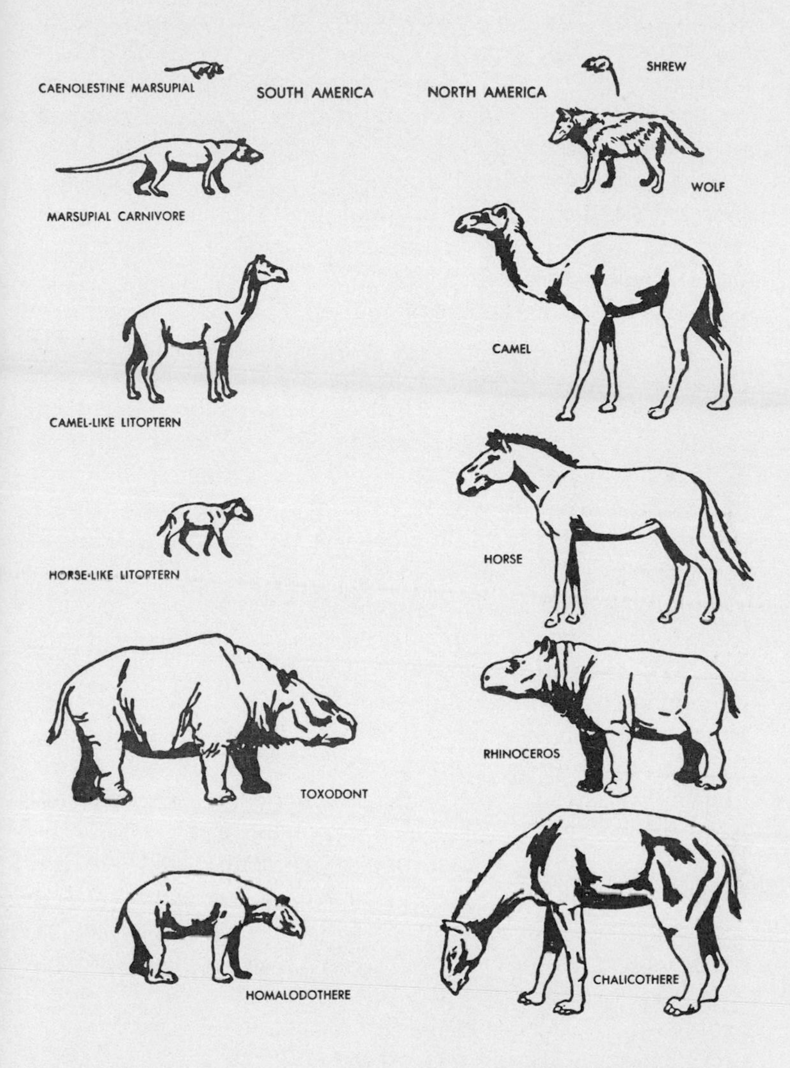

Fig. 32. Some convergent types among North and South American mammals, mostly extinct forms. All are drawn to the same scale.

it would be a stupid anatomist who could not distinguish the two by any single bone or organ in the body. Similarly in South America the extinct macraucheniines may have looked only vaguely like camels but must have had rather similar ways of life. Some of the extinct proterotheres must have looked extraordinarily like some horses, but, as in the case of the thylacine and dog, no single tooth or bone of a proterothere is really exactly like that of any horse. The radiations produced similar results because they represent the development of similar opportunities. They did not produce the same results because they were opportunistic. The opportunities were seized not in the hypothetically most effective ways, not in uniform ways, not, in short, according to plan, but as best they might be, departing from groups already distinctive in character and developing by selection of random mutations which were not the same in the different stocks and not precisely what was needed for mechanical perfection.

Similar phenomena are seen when the evolution of particular organs or structures, rather than that of organisms as a whole, is considered. For instance the basic forelimb of the land vertebrates, five-toed and adapted for walking, has undergone remarkable radiations among later animals. Among other things, it has become a one-toed, tip-toe running apparatus (in proterotheres and horses), an arm and hand (in you and me), a webbed paddle (in duckbill dinosaurs, otters and others), a flipper (in ichthyosaurs, whales, etc.), a wing (in pterodactyls, birds, and bats), and so on. Again recurrent development of the same opportunity is found, and again the opportunism is made evident by similarity without identity and by multiple solutions for the same problems. Throughout, the limb remains essentially the same. It was developed into all these different forms departing from the basis of inheritance from an ancestral type common to all. The forelimb is thus said to be *homologous* among all these vertebrates in spite of such striking differences in structure and function. This common basis and homology have not prevented multiple solutions as an opportunity was seized at different times, by groups with somewhat different forelimbs and with different genetic systems and mutations. Wings, for instance, have developed three times from terrestrial forelimbs (see fig. 33), in the reptilian pterodactyls, in the birds, and in the mammalian bats. In each case the same functional problem, in each case a solution achieving the same end, but in each case a decidedly different structural way of developing a wing.

This sort of opportunistic development of multiple solutions is even more striking when the element of an ultimate ancestral structure in common is lacking. Then the structures developed for the function in question are not homologous but only analogous. Their differences are commonly more radical than in the case, for instance, of the bird and bat wings with their homologous bones, and yet essentially identical functions may be served. Insect wings (fig. 33), as compared with the wings of flying vertebrates, are a good example.

The development of the same sort of evolutionary opportunity by different groups of organisms produces the evolutionary phenomena known as parallelism and convergence. There is little fundamental difference between these two. In parallelism, groups already adaptively and structurally similar independently undergo changes in the same direction (involved in seizing the same sort of opportunity for a modified way of life) within genetic systems originally identical. They thus both change and yet continue to be about equally similar. Such groups are rather closely related to begin with and have not only a structurally similar basis on which change occurs but also similar mutations in similar genetic systems as a means of change. Mutations, like structures, are likely to be more nearly alike in closely than in

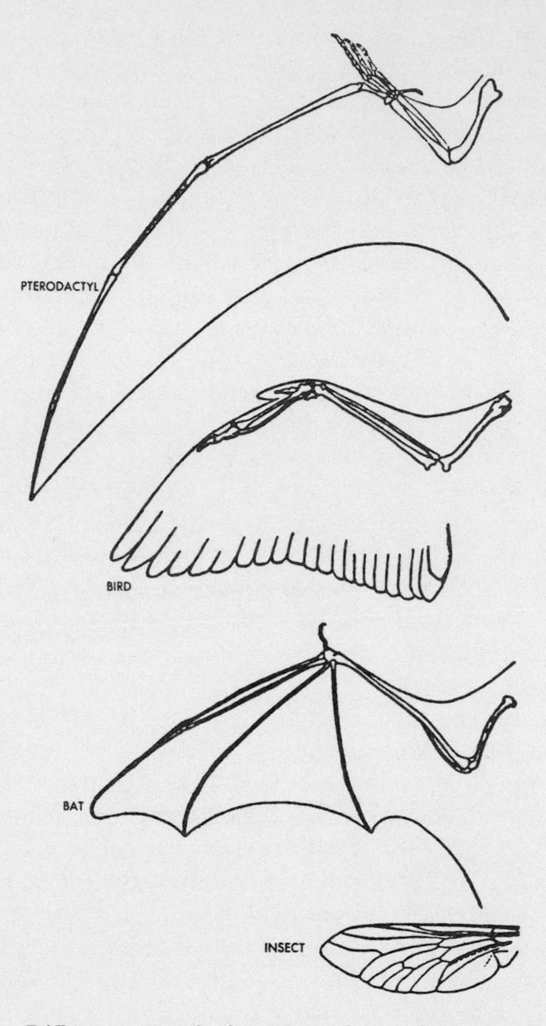

Fig. 33. Different sorts of wings. Pterodactyl (flying reptile), bird, and bat wings developed independently from the forefeet of land-living vertebrates: the bones in them are (largely) homologous; the wings as such and their functions are analogous. The insect wing has a totally different origin: it is analogous only to the other wings shown.

distinctly related animals. It is not surprising that this process may produce forms that are almost, although perhaps never precisely, identical. The gryphaeas, mentioned in discussing oriented evolution, are a good example. They represent several parallel developments from the same continuing oyster stock (*Ostrea*). The end results are in each case so similar that they are commonly considered as collectively forming a single group, *Gryphaea,* although they can be distinguished and do have separate origins within *Ostrea.*

In convergence, there occurs the same sort of opportunistic development of one way of life (or closely similar ways) by different groups, but in this case those groups are dissimilar (or less similar) in adaptive type and genetic systems to start with. Trend toward greater similarity of adaptation involves increasing or converging functional structural characteristics. The groups may be nearly related or may be only very distantly related. Especially among more nearly related groups, the convergence is likely to involve homologous structures, as is also true of parallelism. Homologous structures, having similarities due to common origin, are likely, in opportunistic fashion, to lend themselves to modification for serving similar new functions. On the other hand, among distantly related groups such homologous structures are less likely to be present and convergence may affect organs developed completely independently in each group. The convergent wings of pterodactyls, birds, and bats arise from homologous forelimbs, although they do not arise in the same way. Insects and birds are so distantly related that any particular homology between their parts can hardly be traced, and yet they may converge, sometimes quite closely. Humming moths and humming birds are so remarkably alike in habits and functional operation that they are often mistaken for each other if seen only from a distance.

Convergence on a grand scale is seen in the comparison of South and North American mammals. The "special types" placed opposite each other in the preceding tabulation are in each case convergent. Here, too, is seen the quite general fact that convergence results from divergence, an apparent paradox that is resolved when the opportunism of evolution is taken into account. Divergence within a group involves the seizing of various different evolutionary opportunities. Another group may diverge in a way involving some of the same opportunities. Members of the two groups diverging in ways involving the same opportunity converge toward each other. (See fig. 34.)

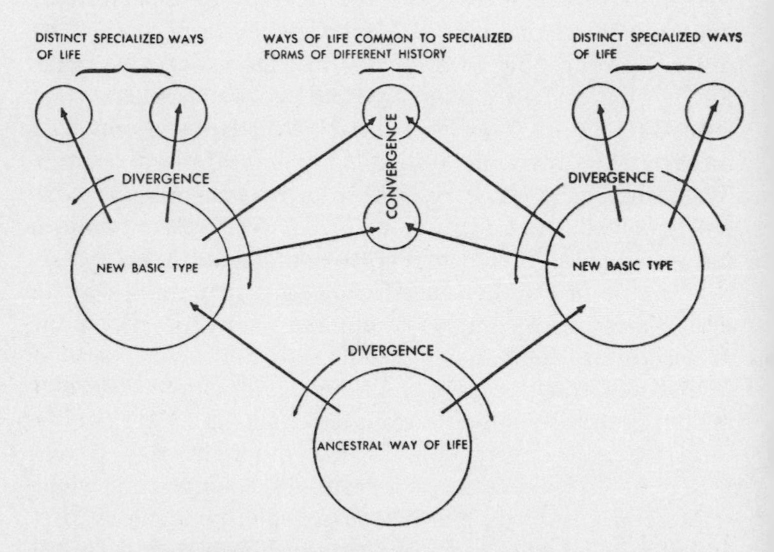

Fig. 34. Relationships between divergence and convergence in adaptive evolution.

This review of some of the opportunistic elements in evolution reinforces the evidence, seen also in discussing oriented evolution, that evolution is neither wholly orderly nor wholly disorderly. It certainly has no grand and uniform plan, nor any steady progression toward a discernible goal. On the other hand, it shows continued trends and a neat interlocking of the various sorts of organisms so that they interact systematically and fill out the possible ways of life, many of which they have themselves, by merely coming into being, created. The history of life is an odd blend of the directed and the random, the systematic and the unsystematic.

As a final remark on this topic, it is to be noted that evolution is not completely systematic even in its opportunism. Huxley's barrel of life, once packed full, does not necessarily remain so. Apparent opportunities are not always exploited and gaps are not invariably soon filled. The rule that all of life's opportunities tend to be followed up also has exceptions. The extinction of the dinosaurs long preceded the reoccupation of most of their ways of life by the mammals—and not quite all seem yet to have been reoccupied. Ichthyosaurs had been extinct for many millions of years before dolphins and their kin seized this opportunity. There is no evident reason why the way of life of the once so abundant ammonites could not be followed now by equally abundant groups for which the seas may be searched in vain. Many types have become extinct and left open a way of life, an opportunity which could not immediately be seized because no other group had yet a structural basis and a supply of mutations appropriate for the change.

Why, then, the extinction? Here we touch on mysteries of life and death, not of individuals but of whole populations, which must also have some special discussion.

XIII. RACIAL LIFE AND DEATH

Races are born, rise to a period of virility, decline, and die. So do individuals. The parallel is tempting. It can be carried further, so that one may speak of "racial youth," "racial maturity," or "racial old age." Even "racial adolescence" may be mentioned, and here the parallel should begin to give pause. Adolescence is the period of maturing during which sexual fertility begins and body growth is essentially completed. Surely races have no period when they become sexually fertile; if not fertile from birth they could never mature. We are in danger here of pushing the analogy too far. Instead of using it only as a sort of metaphor to describe one phenomenon in terms of another quite distinct phenomenon, we have been in danger of assuming that the two phenomena really are the same sort of thing and that we can find in one an equivalent of all the features of the other. That is the trouble with analogies; we may come to think that they will help us to interpret and explain phenomena which, in fact, they can only help, imperfectly and unrealistically, to describe. This confusion has entered into several of the sciences and perhaps most insidiously into the life sciences.[1] Evidently we should not assume that races, because they become old, also become senile, or that when they are young they are virile.

Each normal individual organism has a life pattern which is characteristic of its species and which it follows without strong deviation, barring accident or pathology. Growth

1. One of the most pernicious is the analogy between organism and society. Some further reference to this will later be necessary.

continues for a typical period of time to a typical size as limit. Sexual fertility begins at a characteristic time, continues over a period variable within rather definite limits, and then ceases. Degeneration begins to get the upper hand at some time during the period of maturity and leads to decline, senility, and inevitable death. The potential span of life seems to be very rigidly set by the nature of the organism. All the great strides of medicine do not seem to have raised the potential life span of man by one minute; they have only made it possible for more men to come nearer to realizing their potential spans without being cut off earlier by disease or by avoidably rapid degeneration.

Contrary to an opinion sometimes expressed, races, or groups of organisms in general, do not seem to have any such life pattern which appears as a common element in their histories. Still less do they seem to have an inherent growth pattern or metabolic system which brings them to maturity at definite times and which dooms them to death from the internal ravages of old age, even if violence and disease have passed them by. How extremely varied their life histories may be has already been rather fully discussed and exemplified earlier in these pages in treating the record of their rise and fall. It was seen that most of them do have in common a period of exceptionally high over-all evolutionary activity of various sorts and one, usually near the same time, when they are particularly widespread, varied, abundant, or successful. It was, however, seen that in various groups these periods do not tend to correspond, that they may come early, middle, or late in the history of the group, that they may be long or short, and that they may be single or multiple. The regularity implicit in the evolutionary "Epacme," "Acme," and "Paracme" of Haeckel, the "Typogenese," "Typostase," and "Typolyse" of Schindewolf, the racial "youth," "maturity," and "old age" of some other writers, simply does not exist in this broad picture.

The periods of exceptional evolutionary activity which appear so generally but so irregularly in the histories of groups of organisms are usually ushered in by times of rapid expansion and adaptive radiation, and in most cases they finally taper off with the elimination of many lines of descent and the rather sharp differentiation and narrow adaptive specialization or channeling of the remainder. There is this much regularity and its phases may be likened to "racial youth" and "racial maturity" if you like and if you are not bothered by the fact that "racial youth" may come late, as well as early, in racial life and may occur more than once. In themselves these inconsistencies show that we are not here dealing with any inherent life pattern such as rules the periods of individual life.

The times of rapid expansion, high variability, and beginning adaptive radiation are the "explosive phases" of evolution, to use an infelicitous phrase that seems unfortunately to be deeply established in evolutionary terminology.[2] From what has already been said, it is possible now to interpret these without much further discussion. They are periods when enlarged opportunities are presented to groups able to pursue them.

The opportunities may arise when progress of the group itself brings it to the threshold of a new career: the reptiles evolved to the point of independence from water as a living medium and immediately—as such things go in the earth's slow history—burst into a Virenzperiod (in the Permian) in landscapes earlier barren of vertebrate life. Opportunity may come as an inheritance from the dead, the extinct, who be-

2. Cloud has suggested "eruptive evolution," which is perhaps preferable but not completely satisfying, and Rensch suggests *Virenzperioden,* which would be excellent were it not subject to possible confusion in the individual life analogy. I have suggested "episode of proliferation," but this admittedly clumsier and less picturesque term does not seem to be catching on.

queath adaptive zones free of competitors: Jurassic Virenz for ammonites followed extinction of all but one family, perhaps all but one genus, of Triassic ammonites; early Tertiary mammalian Virenz followed mysterious decimation of the Cretaceous reptiles. Expansion of one group of organisms may mean multiple opportunities for another: spread and diversification of land plants brought on "explosive" evolution of insects and other land animals. Geographic shifts may bring the opportunity of new lands to exploit: precarious island hopping to Australia gave the marsupials a whole continent to occupy.[3]

A quieter period ensues when the basic radiation has been completed, when, so to speak, the opportunities have been parceled out and each has been claimed by some line of populations descending from the expanding group. Lessened activity here betokens not failure or decline but the prosperous enjoyment of a completed conquest. Change continues but mainly in two ways: the sharpening and increasing of adaptation to the way of life embarked upon, and response to changes made necessary in that way of life as time goes on —for no environment is absolutely static. Eliminated lines from the original radiation consist, as a rule, of those so nearly like the surviving lines that they would be in competition with them; these are the *fehlgeschlagenen Anpassungen* of Abel. This, too, is an aspect of the principle of multiple solutions in evolution, discussed from another point of view in the last chapter. In their early phases, adaptive radiations often produce multiple solutions which are simultaneous or closely successive and which occur among animals that are, or come to be, associated with each other. Then one solution may pragmatically prove to be superior to the others, as its possessors displace the rest. Thus in the first radiation of the mammalian carnivores, multiple solutions of the prob-

3. The contemporaneous record of this "explosion" has not been found, but its results leave no doubt that it occurred.

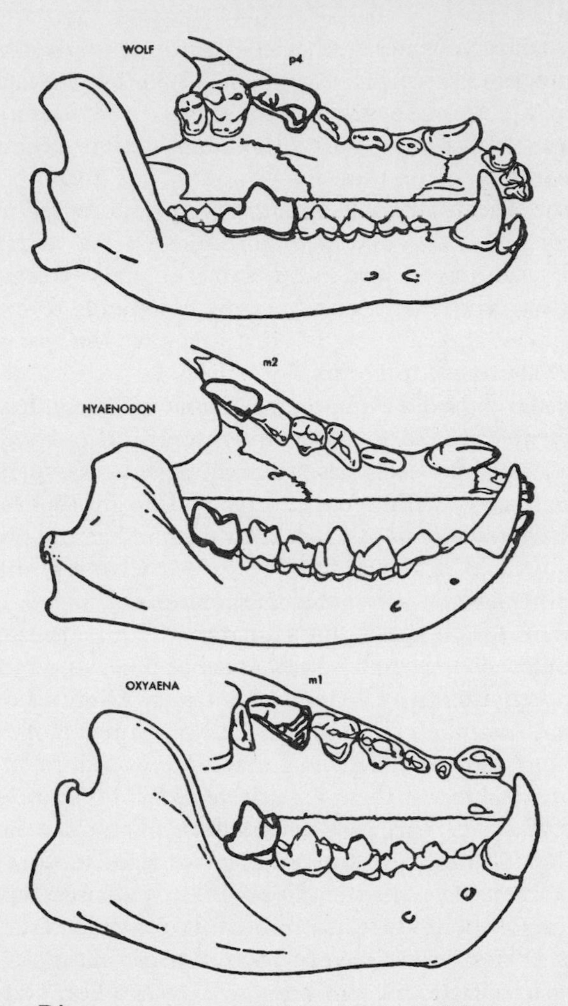

Fig. 35. Divergent, opportunistic formation of flesh-cutting (carnassial) teeth in carnivores. All serve the same function, but different teeth in the series are involved. All occur in the early Tertiary radiation of carnivores, but the type still present in the wolf was most successful and is the only one in living carnivores. (Modified from Matthew.)

lem of developing flesh-shearing teeth occurred (Paleocene and Eocene creodonts). One finally became universal (that of the Miacidae and the fissipeds derived from them) and its possessors later expanded and by their own radiation displaced all other carnivores.

The discussion has been simplified and generalized, but it may suffice to show that these phenomena of racial life history need not involve any mysterious or inherent forces and that they fit well into the general ideas of the evolutionary process that are being developed here. The last phase of life, that of its end in death, is still more important for the problem of the meaning of the history of life, and its study has greater difficulties and uncertainties.

Is extinction the fated end of all groups of organisms, as death is of each individual organism?[4] There are in the world today a large number of types of organisms that arose so long ago that the great majority of their early contemporaries have long been extinct. Some have, indeed, changed enough so that zoologists can apply different specific or generic names to the ancient and recent forms, but in the cases here considered these changes are trivial and represent no essential difference. Among these Methuselahs (see fig. 36) are some already mentioned such as *Sphenodon* (surviving almost unchanged from Jurassic to Recent) or the opossum (Cretaceous to Recent). The invertebrates, older to begin with and generally slower in evolution, provide still more striking examples. The little sea shell *Lingula* is amazingly like its Ordovician ancestor of 450,000,000 years or so ago, and an oyster of 150,000,000 years or more in the past would look perfectly familiar if served in a restaurant today.

4. Discussion of the problem of individuality and of the "immortality" of the ameba, which reproduces by fission, would here be a tempting bypath, but it seems hardly necessary to follow it. Let us agree, in order not to linger over nonessentials, that the ameba as an individual dies when it divides and that the two offspring are new individuals.

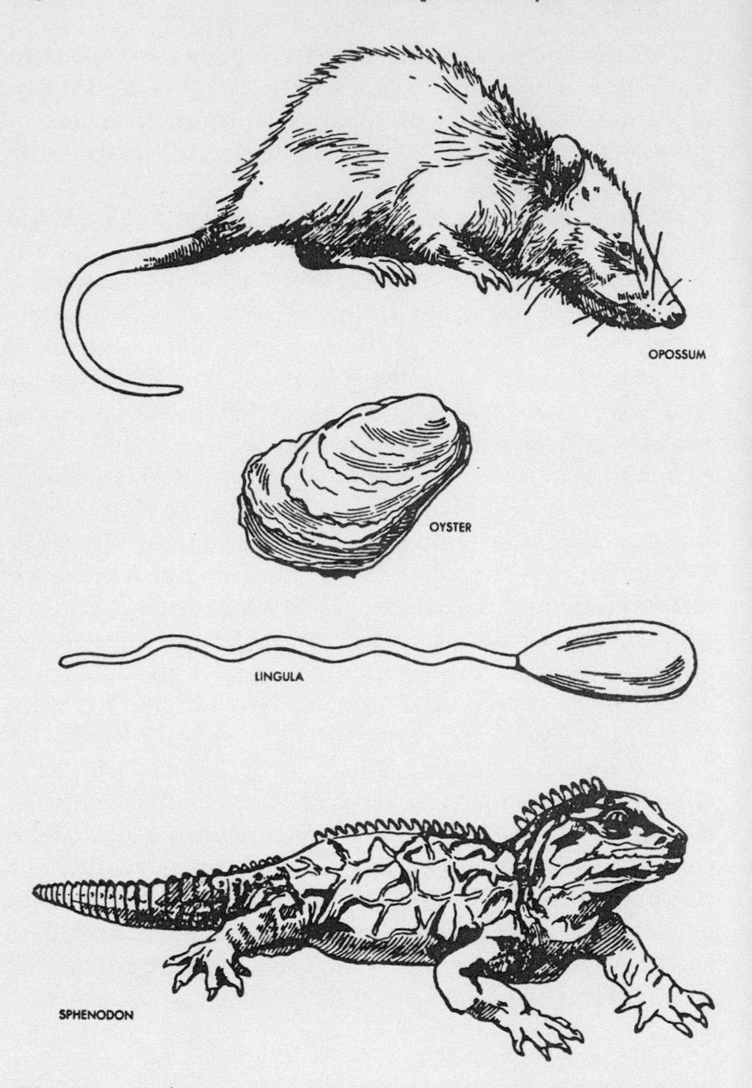

Fig. 36. Some of the "immortals": a few of the living animals, evolution of which has been practically at a standstill for tens of millions of years.

If there is a secret of racial immortality, these animals certainly must possess it. The essentials of the secret are not really very esoteric: when it arose each of these animals was well adapted to a particular way of life which has ever since continued to be possible and the requirements for which have never changed. Some of the offspring of these immortals have embarked on ways of life which did disappear or the requirements for which did change, and these branches have become extinct while the parent stock lived on. From the opossums arose the extinct carnivorous marsupials of South America and from the oysters arose the extinct gryphaeas.

Within these broad conditions are others that do make the problem considerably more complex and not yet fully understood. All environments have undergone some changes, however slight, and the immortals must have met these without need for changing themselves. Mutations away from the original and surviving type have certainly occurred, even with frequency, and these must have been rapidly and surely eliminated by special combinations of factors in the interplay of population and environment.

There are some general features common to many, at least, of the immortals and related in some way to such special requirements for survival. Most of them are relatively small animals; large animals make extra demands on space and on food which may not be met during all fluctuations of the environment. Most of them have rather large local populations where they do occur; small populations are especially liable to accidental elimination and also to chance effects of random mutations away from the established type. Most of them have some latitude in adaptability of individuals and are not rigidly restricted to one variety of food or to a very narrow environmental range; such restriction may be fatal when casual fluctuations in the environment occur. Most of them live in conditions which are not extreme or unusual as to climate, vegetation, and other environmental factors:

in the relatively constant waters of the sea, in temperate forests, and so on, rather than in the Arctic, in deserts, on high mountain peaks, or in other extreme environments. The environments of the immortals are, as a rule, those of longest, most continuous existence. Most of them were progressive animals when they arose and had then evolved rapidly into new adaptive types; they were the first to seize opportunities as they arose and they adapted so well to these that later types did not displace them.

It is now clear, however, that specific adaptation to a stable environment or individual adaptability to a continuing range of environments is not sufficient for evolutionary equilibrium or, as J. S. Huxley calls it, stasigenesis. Populations of the slowly evolving or apparently non-evolving groups have normal genetic variation and are also subject to mutation, which, being random with respect to adaptation, is usually inadaptive. They would therefore tend to lose adaptation and hence become extinct or evolve differently if there were not some positive force maintaining evolutionary stasis. That force is natural selection, which can produce directional, adaptive change but even more commonly impedes change from an already adaptive condition by opposing extreme variants from the mean. In this aspect natural selection is called stabilizing, normalizing, or centripetal.[5]

The fact that extinction has not occurred for these animals during their exceptionally long histories does not permit the conclusion that their extinction never will occur and that they are, in literal fact, immortal. Since the Cambrian, at

5. This effect of natural selection was in fact among the first to be demonstrated unequivocally in a wild population: H. C. Bumpus, "The Variations and Mutations of the Introduced Sparrow *Passer domesticus*" (*Biol. Lectures Marine Biol. Lab. Woods Hole, 1896–1897* [1896], 1–15). It has been discussed at length especially in Schmalhausen's book, previously cited, and in I. M. Lerner, *Genetic Homeostasis* (New York, Wiley, 1954).

least, and probably long before, no extreme or permanent physical change has affected the whole earth. Some sorts of environments in the Cambrian and others developing since then have persisted without essential change. Continuity of basic physical conditions may persist for many millions of years or may end tomorrow. It is incredible that it should really continue forever. The sun's energy is finite and must someday reach an end and with it life on the earth must cease. Collision with other celestial bodies, extremely improbable in such limited time as man conceives, is eventually probable in the endless time of "forever." Man himself seems terrifyingly near the knowledge of how to destroy life on this planet, and terrifyingly far from the self-control necessary to avoid using this knowledge. It must be concluded that material racial immortality is impossible.[6]

Even though extinction is certain for every form of life, the forms we have been discussing have a sort of conditional immortality, which has carried them up to now and which may well continue as long as life on the earth remains possible under conditions not too unlike those of the last billion years or so. Such ability to survive is exceptional. The vast majority of all the multitudes of minor sorts of organisms that have appeared in the history of life have either changed to forms distinctly different or have disappeared absolutely, without descendants.

Among racial life spans, every range has existed from types that arose hundreds of millions of years ago and are still going strong down to types whose geologically brief moment is no more than a few thousand years or so. The differentiation of a distinctly new type of organism normally takes

6. Effective transfer of earthly life to another planet is still a wild fantasy in the light of present (1966) knowledge and technology, and in any case would only delay, not evade, eventual extinction. Possible independent presence of some form of life elsewhere in the universe is not relevant here.

time on the order of hundreds of thousands of years, at least, so that much shorter life spans for such types are improbable. Within any given group of organisms the span of survival for its various sorts or subtypes is extremely variable. The whole range, from the kind of conditional immortality discussed above down to a minimum, may occur in a single group, as it does, for instance, in the pelecypods or bivalves, the clams and their host of allies. Yet within each group there seems to be an average span of racial life which is characteristic for that group and which differs from one group to another. Thus the genera of mammalian land carnivores that have become extinct lasted, on an average, for a period on the order of 8,000,000 years, although some endured for up to 20,000,000 years, more or less (middle Miocene to Pleistocene), and others disappeared almost as soon as they arose. Extinct pelecypod genera[7] had an average span on the order of some 80,000,000 years, perhaps ten times as long as that of the carnivores. Some lasted more than 200,000,000 years before becoming extinct and others again were barely differentiated before they disappeared. Note that the maximum span—for genera that did become extinct; the "immortals" are excluded—is about ten times that for carnivores, as is the average span, and that in each case the maximum is about 2.5 times the average.[8] It is also characteristic of each of these groups, and probably of most other groups of organisms, that short life spans are much more common than long: the great majority of genera became extinct after lengths of time nearer the minimum than the maximum for the group as a whole. Extremely short spans, those quite close to a minimum, are, however, less common than those of somewhat greater duration. Most genera tend

7. Genera construed broadly and not the extremely split genera of some recent classifiers.

8. The figures are taken from my *The Major Features of Evolution*, previously cited.

to have about the same period of survival, characteristic for each group, and those with very short or with moderately to very long relative durations are less common.

There is, then, a rule of extinction for various groups of organisms, and while the figures differ greatly from one group to another there are regularities common to some at least of these groups. Although we failed to find good evidence of a widespread racial life cycle pattern, here is a suggestion of such a pattern for death. The pattern is real, but a little further study of its causes and meaning shows that it is not inherent in the organisms alone, not a fate implanted within the racial tissues as inevitable old age and death are implanted in the tissues of individuals, but is again a phenomenon explicable only on the basis of the complex material interaction of populations and their environments.

For one thing, the pelecypods include a considerable number of the "immortals," and the carnivores include none (so far as we can judge without knowing their future). Some pelecypods still living have been in existence twice as long as the maximum for those that did become extinct. No living genus of carnivores is as old as the maximum span for extinct genera in this group. The pelecypods have available environments within the sea which have undergone no great change through hundreds of millions of years. Some of them have been able to seize this opportunity for indefinitely prolonged survival. Environments available to the land carnivores have changed much more rapidly and more radically. Correspondingly, they have had shorter spans and no opportunities for indefinite continuance of any one type. Aside from these differences, pelecypods have coped, or failed to cope, with the slower changes of the sea and carnivores with the faster changes of the land in accordance with the same principles of adaptation and organic stability and modification.

The figures for extinction have included two different

things. If an evolving population of organisms changes so much that it becomes something distinctly different, to be given a different name and considered something new, then the old, ancestral type is gone, extinct, and yet it also lives on in altered form. Extinction in this sense merely reflects the rate of progressive change. A short life span for any one type means not that death soon overtook it but that life rapidly carried it on to other forms of being. Total figures for extinction, such as those given for pelecypods and carnivores, include many cases of this sort and thus are in considerable part reflections of rate of evolution, which is clearly much more rapid in carnivores than in pelecypods.

Quite commonly when progressive change evolves a new type of organism, it arises from only part of the ancestral population. The old type also continues to exist, as in the now familiar example of *Ostrea* and *Gryphaea*. There are thus really three different cases. A line of descent may give off a distinctive branch but itself continue without much change; extinction of the parent stock has not occurred in any sense of the word. Or, second, it may change as a whole into a new type, or several; there has been an extinction of a particular type of organization but not absolute extinction of a line of descent. Third, it may cease to exist, its last representatives dying without issue; this is extinction absolute and unqualified, better labeled as termination.

It would not be quite right to say of the second of these cases that no extinction has occurred. Extinction of type is obviously different from absolute ending of a line of descent, and it is clearer to give different names to the two, but in both cases a sort of organism has disappeared from the world. Now in dealing with the record of the history of life we saw that broad types of organization tend to persist indefinitely. Extinction for them is not the rule but is decidedly exceptional. It is doubtful whether any animal phylum has ever become extinct. Of eight classes of vertebrates only one became extinct. These major grades of organization represent adaptation of a broad sort to general ways of life, such as (for Protozoa) microscopic floating life in a liquid medium (with a few other almost equally general qualifications). The possibilities for following such a general way of life have persisted, and so have the corresponding general types of organization. A more specific type of organization or, within a larger group, a single line of descent through continually interbreeding and reproducing populations, follows a way of life integrated with more particular environmental conditions. These may persist indefinitely, as they have for the "immortals" among animal types, but more commonly they tend to change over longer or shorter periods of time. Survival then demands changes in the integration of organism and environment and this, if it exceeds a certain and commonly rather narrow range of tolerance, requires evolutionary changes in the organisms. They either make these changes, in which case their former sort of organization becomes extinct, or they fail to make them and their absolute extinction ensues.

A type of organization normally persists as long as its adaptive relationships continue to be possible. Survivals of all phyla, of most classes, of a few genera, but of no species, are all exemplifications of this rule. Its obverse is exemplified by extinction of all species, of most genera, of some classes, but of no phyla.

This view of extinction needs further examination and particularization, but first it may be desirable to take brief notice of some opposing views. One of these views is that extinction is caused, sometimes or normally, by the continuation of an inherent trend that is not necessarily adaptive in nature or orientation. This possibility has already been considered in previous discussion of oriented evolution, with the conclusion that there is no strong evidence that such a thing ever did occur. It may be noted in passing that this view

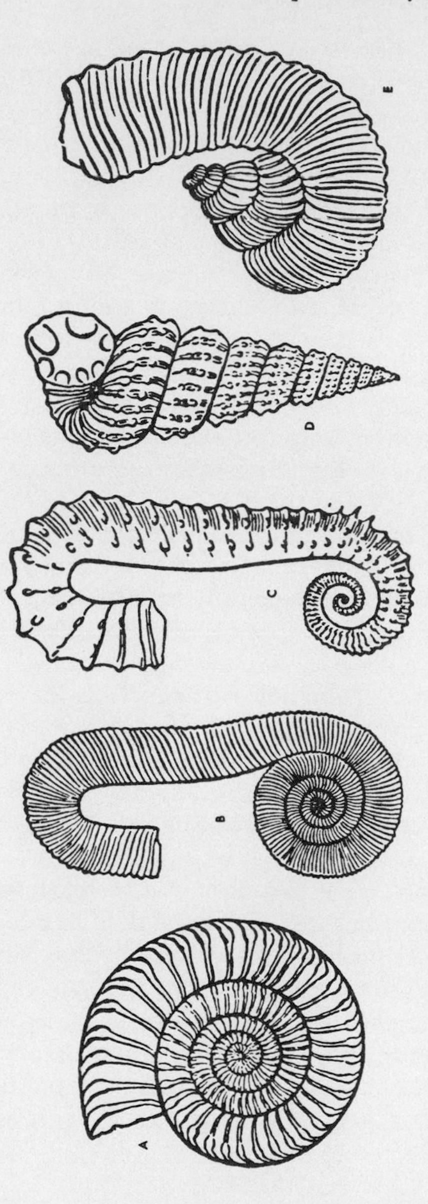

Fig. 37. A few ammonites. A, the "normal" or usual, simply coiled type. B–E, some of the unusual types considered by some students as indicative of "racial old age." (Data from Boule and Piveteau.)

(and also the next to be considered) is most commonly held by those whose philosophy of evolution is vitalist, finalist, or both. I cannot help finding it very strange that anyone should think that an entelechy or vital urge should orient life toward death, or that the supernal plan followed by life has death as its goal. If—the finalist reply—these are only side lines that missed the goal, is it not impious to impute such fumbling to the Planner?

Another view is that groups of organisms become old, just as individual organisms do, that senility sets in and extinction follows from senile degeneration. Some suspicion of this forcing of the analogy has already been expressed. Examination of the supposed evidence for this phenomenon tends only to deepen the suspicion, or to convert it into a conviction, that the phenomenon does not really exist. The example most often given (first, perhaps, by the great American paleontologist Hyatt in the 1890's) involves the ammonites, a group that has already often been mentioned in these pages because it lends itself so well to exemplification of evolutionary processes and has long been studied by a large number of the ablest investigators of these processes.

The ammonite group as a whole, consisting of marine invertebrate animals with chambered shells separated by more or less complexly plicated partitions, persisted at least from early Devonian to late Cretaceous, a span of more or less 310,000,000 years. The typical or commonest form of the shell was a spiral coil in a single plane, somewhat like a heavy line coiled on a deck, except that later coils usually overlapped the earlier. In the Mesozoic some peculiar variations began to appear. (See fig. 37.) The coil might be open so that succeeding coils did not touch the earlier, or only the earliest parts might be coiled and the later perfectly straight; early coils might be followed by a straight segment and then by another coil turning back toward the early part; the plane might be abandoned and the spiral grow upward in an asymmetrical, snaillike way, or growth might even be very irregular with no clear spiral.

The claim is that these irregularities are signs of senility, that growth was no longer under proper control because of lessened vitality in racial old age. In fact such types occurred, and periodically recurred, from the Triassic onward, in other words during the whole latter half of the long history of the group. They were common during times when the group was most widespread, abundant, actively evolving, and obviously eminently "virile" and successful. A family particularly characterized throughout its history by such aberrant growth forms (the Lytoceratidae) was one of the longest lived of ammonite families, persisting well over 100,000,000 years through the Jurassic and Cretaceous. Moreover, forms with perfectly normal plane spiral shells continued abundantly right up to the end and then became extinct along with all the others. The paleontologists who have seen evidence of senility in this history have been victims of the pathetic fallacy. They have mistaken mere unusualness of shape, which clearly was, at worst, no impediment to vigorous development and good adaptation, for the deformities of old age that beset ourselves as individuals.

It is true that the last members of a group before its extinction are frequently—not always, by any means—extreme in size, horn development, or other such features and occasionally quite bizarre. Of course if a group is changing progressively, its last members are bound to be an extreme for that group; if it had lasted longer, it would have reached new extremes. As a matter of fact, animals about as bizarre as any that ever became extinct are alive today and doing well: elephants are as queer as mammoths; living whales are far bulkier than any dinosaur; the spiral "unicorn" tooth of the narwhal has no equal for strangeness among past animals; it would be hard to imagine anything more fantastic than some insects such as a dynastes beetle or some of the mantids.

To the extent that such developments may perhaps have been involved in extinction, there are at least two concrete and established factors fully competent to account for this without calling in the vague and dubious factor of supposed racial old age. First, strong structural peculiarity usually betokens high specificity of adaptation and limits avenues of possible change, and these are certainly factors making extinction more likely. Second, when a group is already waning and approaching the danger of extinction, its local interbreeding populations eventually fall below the size where random, inadaptive mutations are regularly eliminated without becoming fixed in an undue proportion of the population. This, with accompanying excessive inbreeding also likely in such a situation, may tend to produce bizarre, sickly, or generally inadaptive forms. The process will hasten extinction, but it sets in only when extinction is already approaching. If this be defined as "racial old age," then impending extinction causes "old age," not the other way around.

The general, true cause of extinction seems to be a change in the life situation, the organism-environment integration, requiring in the organisms concerned an adaptive change which they are unable to make. The sorts of environmental changes involved may be extremely varied and the inability of the organisms to make appropriate adaptive responses is also a complex problem. A factor of prime importance is specificity of adaptation or narrowness of individual, nongenetic adaptability. Allusion has already been made to the wide differences shown by animals in this respect, for instance as regards diet. Obviously an animal that can live on a wide variety of different foods is more likely to survive if food in general becomes scarce or if some kinds disappear than is an animal that can eat only one kind of food.

Specificity of adaptation is what is meant in the main by specialization in evolution, and "overspecialization" is a major factor in extinction. It must, however, be noted that every animal that ever lived and all that live now are specialized to some extent: each has special life requirements and none can live in more than a small fraction of the environments offered by the earth. This universal specialization becomes overspecialization only when it fails to meet the pragmatic test of survival, and this failure can occur with any degree of specialization. If a deluge were to cover all the lands of the earth, all land animals would be overspecialized in the face of this emergency. As long as there is land to live on, land life is, for most of its practitioners, a specialization with good survival value and not an overspecialization.

There is a reciprocal relationship between breadth of specialization and likelihood of extinction. Slight environmental change will harass a group of narrow specificity in adaptation and may cause extinction if the group does not have appropriate variation to make an adaptive change. Even if such a change is possible to the organisms concerned, it may be blocked if, for instance, some other group is already firmly established and well adapted in the new way of life which affords an otherwise possible escape from extinction. Larger environmental change will, as a rule, have to occur before extinction threatens a group of less specific adaptation, but in the end any group may be threatened or may become extinct if the change becomes radical enough.

Environmental change has constantly occurred and so, throughout the history of life, extinction has repeatedly befallen groups with specialization such as to be critical in the given emergency. Since the more specialized groups are more liable to extinction, it may seem remarkable that increasing specialization is nevertheless the rule in the histories of most groups of organisms and that a stage is never approached when all surviving groups have the lowest practicable specificity of adaptation. In fact, particular environmental conditions may often remain essentially constant over long periods, the critical emergency may be long delayed or may never arise. Disappearance of grass would have doomed most of the sorts of later Tertiary horses and various other animals, but grass did not disappear.

In the meantime, there is an advantage in specific adaptation as long as its environmental requisites continue. Grass is an abundant food that cannot be effectively used except by those animals specifically adapted to it. There is in evolution a continual balancing of the two advantages: the advantage of increasing specialization in sufficiently stable conditions, the advantage of versatility in changing conditions. A crude but picturesque analogy, which we will pursue no further than its validity for metaphoric description, is provided by specialization in human occupations. A highly specialized artisan such as, say, a glass-eye blower, will thrive as long as his trade is adaptive—there is a brisk demand for his products—and will be better paid than a general laborer. If there were no further demand for artificial eyes or if a cheap, improved type not blown from glass were developed, the glass-eye blower would have to change his trade or starve, while the general laborer would be unaffected.

There is also the point that one specialization may lead to another, and here the descriptive analogy continues to hold. If glass eyes went out of demand, the blower of them might yet maintain his advantage by turning his special skills, say, to the making of glass laboratory apparatus. Specialized animals often similarly avoid extinction by, so to speak, changing their trades. Browsing horses became highly specialized and some reached a point where further and different specialization for grazing became possible. The browsers eventually all became extinct; the grazers did not. Sloths originally specialized for life on the ground and for digging. Some utilized features of this specialization convertible for a very specialized sort of arboreal life. The ground sloths are extinct; the tree sloths are not.

The particular changes that cause extinction when they affect groups unable to adapt to them are extremely varied.

Any environmental change whatever may have this result. Probably one of the commonest initiators of extinction is competition. The environmental change is the development or incursion into a given region of a group of animals whose adaptations and consequent life requirements are essentially similar to those of a group already present. The two groups, different to some degree in history and origin, are necessarily not identical in structure and function. (We have already seen that parallel or convergent adaptation cannot produce complete identity.) In some respects, however slight, one group will have a balance of characteristics advantageous to it. The other group will inevitably either become extinct[9] or develop a new specialization outside the range of its successful rival.

In this process it is an interesting feature that the more specialized group usually has the advantage and is the one to retain its field of specialization. If the less specialized group does not simply become extinct, it tends itself to become more specialized, but specialized in a different field from that of the group with originally higher specialization.[10] This is the process grandly exemplified in the thin-

9. Aside from evidence of this outcome in free nature, it has been established experimentally that extinction always occurs when two species are in competition for food and neither has means for withdrawing from the competition. See, for instance, T. Park, "Experimental Studies of Interspecies Competition. I. Competition between Populations of the Flour Beetles, *Triboleum confusum* Duval and *Triboleum castaneum* Herbst" (*Ecological Monographs, 18* [1948], 265–308). Like so many supposed discoveries in our century, this principle was already clearly discussed by Darwin, although he had no experimental evidence. For a more recent review see G. Hardin, "The Competitive Exclusion Principle" (*Science, 131* [1960], 1291–1297).

10. This was also known to Darwin. One of the narrower aspects of it was given a new name and reviewed in W. L. Brown and E. O. Wilson, "Character Displacement" (*Systematic Zool., 5* [1956], 49–64).

ning out and increasing divergence and specialization of the various lines of descent in an adaptive radiation. It helps to explain why increasing specialization is one of the most widespread features of evolution although it is also one of the commonest factors in extinction.

Extinction by competition is also the usual process in replacement of one group of organisms by another, although we have also noted cases of delayed replacement in which the extinction has some other cause and replacment occurs after, not concurrently with, extinction.

There is another process of specialization and competition that may have had considerable importance in evolution. Competition commonly occurs not only between but also within groups of organisms. Among many polygamous mammals and birds, for instance, the males engage in overt, active competition for the females. Specialization to meet competition within the group then often involves evolution of weapons or ornaments, such as antlers in deer or trains in peacocks. Within most species there is some degree of competition, generally passive, for food or for living room, and this too may involve specialization. It is probably a factor in the frequent tendency toward larger size in evolution, for the larger animals of such groups commonly do have some advantage over the smaller in competition with their own kind. Such specialization for intragroup competition may also be of some use in intergroup competition, as would generally be true of larger size and as seems to be true to a limited extent, at least, of deer antlers. In other cases, such as that of the peacock's train, it is hard to see any possible advantage for intergroup competition and there may be a definite disadvantage. In all cases, specialization for intragroup competition, simply because it is not developed with reference to survival or successful intergroup competition by the species as a whole, is a possible disadvantage to the species in its external relationships. It may well determine the loser when intergroup competition does occur. The suggestion has also been made that it may even be carried to such a point as to cause extinction without intervention of any competitor or environmental change, but for this I can see no evidence; the sabertooth canines are sometimes given as an example here, too, and we know that they certainly did not cause extinction by their overgrowth—it is also extremely improbable that they were specializations for intragroup competition.

Particular cases of extinction, other than those evidently due to competition, are usually hard or impossible to explain in detail. It is not that there is any serious doubt about the general cause, but that possible particular causes present an embarrassment of riches and many of them can leave no clear trace in the record of earth and life history. We do not know just why horses became extinct in the Americas around the end of the Pleistocene, not because the event is inexplicable but because there is no conclusive way of choosing among possible explanations. Change of climate or topography, decreasing food supply, spread of poisonous plants, incursion of new beasts of prey (such as man, in this case), outbreak of epidemic disease, and many other possibilities come into consideration in such cases. Some can sometimes be shown to be improbable, but elimination can seldom be carried down to one possibility, and indeed in many cases it is probably an accumulation of changes and not one alone that initiates extinction. The process is also cumulative, because once a group has begun to decline it will sooner or later reach a critical point beyond which the final elimination may be produced by factors that would have had no fatal influence while the group was abundant. Among factors of this sort are seasonal fluctuations in numbers, unchecked inadaptive mutation, excessive inbreeding (with resulting undue proportions of disadvantageous or lethal homozygous recessives), or geographic restriction—so that purely local conditions or catastrophes may affect the fate of the whole group, as they cannot in a group that is widespread.

It is not intended to stress environmental changes as if they were in themselves the cause of extinction. No such change can cause extinction if the organisms affected also change in an appropriate way. On the other hand, we have seen that organisms can survive indefinitely without changing in any essential, or indeed just *because* they did not change. It is neither the change in environment nor the lack of change in the organisms that causes extinction: it is the two together. Much of the long scientific discussion of theories of evolution has turned on the question whether evolution is fundamentally autogenetic (or autonomic) or ectogenetic, that is, whether the driving factors in the history of life have been internal to the organisms or external to them (environmental). It has become clear that this is a false issue. The truth is not in one or the other of these alternatives but in a third which combines the two and yet has essential features of its own.

XIV. FORCES OF EVOLUTION AND THEIR INTEGRATION

The course of evolution is marked by changes in structures and functions as these appear in individual organisms. The evolutionary forces act, of necessity, on individuals, but their effects are not achieved within single individuals. These effects work out within associated groups of individuals, populations, and over the course of successive generations. A key to the process is, then, that evolutionary changes must arise in or must, at least, be transmitted by whatever it is that passes from one generation to the next, by the heredity of the group. The association of individuals into populations that is important in this process must therefore be reproductive association. The group that actually evolves is a group linked together by exchange and transmission of hereditary factors, a population of individuals that reproduce, usually by interbreeding, over the course of generations. It is true that the survival of individuals in a community and their success in producing new generations may be strongly affected by nonreproducing members of the community (such as the unsexed workers in insect societies and the older or otherwise nonbreeding members of human society). In the special circumstances of communal life the nonreproducers too may thus have an important role in evolution. It is nevertheless fundamental that evolutionary change consists basically of change in heredity and that this is developed in reproducing populations.

Heredity is carried for the most part by genes, which are segments of replicating molecules (of DNA) which occur within chromosomes, which also usually are replicated faithfully whenever one cell divides into two. These chromosomes, with their included genes, are passed on from one generation to the next, usually within specialized germ cells from which individuals of the new generation develop. Their development is mainly controlled or determined by the genes in chromosomes received from parents, and under this control it follows a pattern that produces characteristic body forms and physiological systems similar to those of the parents.

Heredity is, on the whole, a conservative factor tending to keep succeeding generations within a common pattern. The acorn produces an oak similar to the tree that produced the acorn, except in unessential details, and the egg produces a chicken essentially like the hen that laid it. We know, nevertheless, that offspring are never *precisely* like their parents, and since evolution is a process of change these differences are of special concern to us here. Such differences may arise in three ways, and, in the vast majority of cases, in these ways only.

In the first place, the development of the individual is affected not only by the inherited growth determinants but also by the conditions under which growth occurs. Like most of the processes important in evolution, individual development turns out on closer examination to be neither purely autogenous nor purely ectogenous but a combination of the two. No two oak trees, even if grown from twin acorns, have exactly the same size, shape, placing of limbs, number of leaves, etc. Greater or less differences in the environment (for instance in soil or local weather) affect them as they grow, no matter how near each other they may be. Such effects occur to markedly different degrees in different sorts of organisms. When seeds or eggs are simply cast adrift to develop as best they may, their development is strongly affected by the environment. Plants as a whole have much more of such environmentally induced variation than do animals. At the other extreme, probably the most important feature of the mammals is that in them this sort of variation is reduced to a minimum. Development within the nearly uniform environment of the womb followed by maternal care and milk feeding ensures reproduction about as true to type as may be, although of course even here absolute uniformity of the conditions surrounding growth is not obtainable. Grosser changes are still caused by disease and accident, and different activities of the organisms themselves cause change in them, as, for instance, exercise may increase the size of a muscle or lack of use decrease it.

Differences between parents and offspring arising in this way are not hereditary. It was long supposed that they were and a whole theory of evolution, to be mentioned later for its historical interest, was based on this belief; but this has now been disproved beyond reasonable doubt. It is, nevertheless, going too far to suppose that such changes have nothing to do with evolution. They are not themselves changes in heredity such as, in a fundamental sense, constitute evolution, but they may have a decided influence on these hereditary changes. Adaptation to local environment during growth may have definite survival value to the individual and affect its passing on of heredity to future generations. Any individual change not itself hereditary may nevertheless strongly influence the individual's ability or opportunity to reproduce, and in this indirect way will affect the course of evolution of the group as a whole. Individual flexibility or adaptability is also itself an essential characteristic that is heritable and that evolves.

A second and, from an evolutionary point of view, far more important source of differences between parents and offspring is in the mechanism of sexual reproduction, which is not quite, but nearly, universal among both plants and animals. (Even the protozoans, which usually reproduce by simple fission of the whole body, normally undergo from time to time a form of sexual reproduction, and even bacteria, which lack sex in the strict biological sense, do sometimes interchange and recombine genetic material.) The offspring receive chromosomes, and hence genes that are bundled into chromosomes, from two different sources, usually an equal number from each in corresponding pairs. It is extremely unusual for two individuals to have identical sets of genes, so that the chromosomes received from the two parents in sexual reproduction are practically always different in some respects. The combination of the two sets in the offspring thus differs from that in either parent. When the offspring, in turn, come to reproduce, each passes on to its progeny not simply the set of chromosomes it received from one parent or the other, but a random selection from both sets.

Thus in continued sexual reproduction through the generations the genes and chromosomes are constantly being reshuffled. Even in organisms with relatively simple gene sets and few chromosomes, the number of possible combinations of different genes and dissimilar chromosomes is astronomically large, far larger than the number of individual organisms in the line of successive interbreeding populations. It is rare for any two individuals to have exactly the same sets of these heredity determinants; in fully sexual reproduction this can happen only when they are identical twins, developed from a single fertilized egg—and such twins, although exactly like each other in heredity, will almost never be exactly like either of their parents. Heredity is essentially conservative in its broad outlines, but this mechanism of shuffling in sexual reproduction makes it almost endlessly varied in detail.

Such changes between parents and offspring are hereditary, subject to continuation of the same sort of shuffling that produces them, and are therefore materials available for basic evolutionary change. Combinations produced in the shuffling process are not necessarily soon lost again by reshuffling. Recombination of genes within single chromosomes is an occasional, not an invariable, process and it is not wholly random. Particular combinations of chromosome pairs, even though produced at random, become more frequent in the population if each sort of chromosome involved becomes more frequent. Identity of all the thousands of genes and of the several or dozens of pairs of chromosomes is exceedingly rare between any two individuals, but similar groupings and identity of certain limited combinations of a few key genes may readily be developed within a large proportion or all of a population.

The role of these changes in evolution is important, but it has limits. New combinations of genes and chromosomes produce new variant sorts of organisms, but no basically new types of organisms can arise and evolutionary change cannot be long sustained, geologically speaking, as long as the genes and chromosome sets remain of the same kinds. Major and long-continued evolutionary changes therefore depend on a third source of difference between parents and offspring: mutations, which are the production of new sorts of genes and chromosome sets. Chromosome mutations may change the number of chromosomes (a change fairly common in plants but less so in animals) or their form and character, for instance by reduplication of segments, or reversal in their position within the chromosome, or their transfer from one chromosome to another. Such differences may or may not have more far-reaching effects than reshuffling of existing genes and chromosome sets, but they too are limited by the fact that the same sorts of genes, the ultimate determinants as far as we know, continue to appear in them. The most basic changes involved in evolution are gene mutations, the rise of new sorts of genes.

Only recently (in fact since the first version of this book was written) has it been discovered what genes are in a concretely material sense. They were earlier defined operationally, by their effects on developing organisms. Even though it was not then known in chemical terms what a gene is, it became well known how in general genes act, what many individual genes do, and even just how they are arranged in particular chromosomes. Now we know that genes are segments of molecules of DNA (deoxyribonucleic acids)[1] that carry what can be figuratively called a coded message by different sequences of four kinds of subunits called nucleotides. A segment that corresponds with the operational gene determines (by indirect processes not needing specification here) the structure of a particular protein, the sequence of its amino acids, synthesized in the cell. Those proteins, mostly enzymes, in turn determine essential characters of the developing organism as well as of its cell physiology after development.

Usually when chromosomes are replicated, their chromosomal DNA and included gene segments are precisely copied. Occasionally they are not, and then a gene mutation has occurred. The mutation is a change in nucleotide sequence, hence in coding for a protein. The various causes and kinds of such nucleotide changes are among the things still not completely understood. More important and more deplorable is the fact that we still know extremely little about the exact structure (nucleotide sequence) of particular genes and still less about how that structure actually determines the characteristics of a given organism as a functional whole. Thus it is that the truly brilliant recent discoveries in nucleic acid and enzyme chemistry have made little contribution to knowledge of whole, living organisms and of their evolution. With only minor exceptions, genetical aspects of evolution can still be best studied and explained in terms of the operational, not the directly chemical, concept of genes.

1. In rare instances genes are in other forms of nucleic acids.

Among the facts of particular evolutionary importance that are known about genes are these: they are inherited as discrete, unblending units but they act in development as interacting and cooperative sets; they do not correspond in a one-to-one relationship with structures of the developed organism above the molecular level, but affect or control the growth pattern from which those structures arise; through this growth control single genes may have effects on various different structures and characters (this is called pleiotropism); one gene or several may modify the action of another; and a single character may depend for its development on the presence and interaction of a number of different genes.

Some of the things known about gene mutation and important for study of evolution are these:

Some genes are very stable, some mutate with considerable frequency; it is probable that all can and do mutate in time. Frequency of mutation may be changed by various influences such as the presence of certain other genes or application of heat, radiation, or chemicals; but with some qualifications the nature of the mutation is not determined by such influences. The effect of a single mutation on structural or physiological characters may be almost imperceptible or may result in obvious and radical changes in the whole body. "Large" mutations, those with the greatest effects on the organism, are commonly, but not necessarily, lethal, preventing development of the fertilized egg or bringing early death to the developing individual. Most important of all, from our present point of view, the effects of gene mutations have no evident relationship to the adaptive status, needs, or general way of life of the organism involved. They are in this respect entirely random. This is true, too, of induced mutations; for instance, the effects of those caused by application of heat have no particular relationship to adaptation to temperature of environment. In an organism already well adapted, almost all mutations will be disadvantageous in the previously fixed way of life, simply because in such conditions any random change is likely to be for the worse. The change may, however, or may not, be advantageous in some other way of life which may, or again may not, be accessible to the organism.[2]

One of the frontiers of modern genetical and evolutionary research is study of the way in which mutations affect different stages in development of the organism and the evolutionary roles of such effects. The different possibilities have been explored in considerable detail and a technical name has been given to each. Some of these names have a certain awesome fascination ("lipopalingenesis," for example) but they need not delay us here. Changes made evident only in later stages of development generally have rather slight effects on final outcome. If they consist essentially of additions at the end of the developmental process, they leave most of that process, as inherited from the ancestry, intact. This sort of change, which is common but far from universal and seldom entirely clear-cut in the evolutionary process, is responsible for very rough approximation of successive ancestral stages in the development of the individual in accordance with the overgeneralized and much abused aphorism of the

2. The serious enquirer should supplement this minimal presentation of such facts of genetics as are absolutely necessary for understanding of the meaning of evolution by reading some modern treatment of genetics as a whole, such as A. M. Srb and R. D. Owen, *General Genetics* (San Francisco, Freeman, 1952), or G. Beadle and M. Beadle, *The Language of Life* (Garden City, Doubleday, 1966). Th. Dobzhansky's *Genetics and the Origin of Species* (3d ed. New York, Columbia University Press, 1951), which takes for granted knowledge of the elements of genetics, is extremely important and is not outdated by later developments in molecular genetics. It has had a fundamental influence on current comprehension of evolution and my own understanding and discussion of the subject, throughout this book and otherwise, have drawn heavily on it.

nineteenth-century evolutionists that "ontogeny repeats phylogeny."

On the other hand, the effects of a mutation may start to operate early in the development of the individual. There is some ground for suspicion, although the point cannot be said to be established, that this is the usual course, that mutations ordinarily affect the whole process of development. Their effects may, of course, be more evident in one stage or another. If those effects are slight but cumulative, they may not be noticeable until late stages and may be supposed to be confined to those stages. In any event, if effects are really confined to late stages, their results cannot be very profound, but if they begin to appear early in development the final results may show radical changes from the parental condition, or may again be only slight.[3] A marked alteration of early course of development with consequent still more marked changes in final structure seems to be the usual mechanism for "large" mutations. This in turn provides a conceivable mechanism for sudden origin of new structural types in evolution. The possibility is evident, but its general significance in evolution is improbable, and it is very unlikely that this is the way in which basic new types usually arise. This will be mentioned again.

The random nature of some kinds of changes in heredity must be particularly emphasized. The shuffling of existing

3. The problems and significance of individual development have been discussed from quite different points of view in two previous Terry Lecture volumes: *Order and Life* by J. Needham (New Haven, Yale University Press, 1936); and *Ourselves Unborn,* by G. W. Corner (New Haven, Yale University Press, 1944). Other interesting studies on the relationships of embryology and evolution include G. R. De Beer, *Embryos and Ancestors* (Oxford, Clarendon Press, 1951), and C. H. Waddington, *The Strategy of the Genes* (London, Allen and Unwin, 1957) and *New Patterns in Genetics and Development* (New York, Columbia University Press, 1962).

stocks of genes in sexual reproduction is, in the main, random. The appearance of chromosome and gene mutations is also largely, although not completely, random and the nature of their effects seems to be altogether random with respect to the needs or adaptation of the organisms and with respect to the direction in which evolution has, in fact, been progressing in the given group. This led some of the earlier students of heredity to think that evolution is really a wholly random process. "Early" here means within the present century, for almost all of our real knowledge of how heredity works has been gained in that time.

In examination of the phenomena of evolution, we have seen that there are indeed elements in these that appear to be erratic and to arise at random. The source of these random effects has now been identified: it is, broadly, the largely random operation of the whole mechanism of heredity. We also saw, however, that many of the phenomena of evolution are clearly oriented to some degree. The history of life has not been strictly random or strictly oriented, but an odd mixture of the two, with one predominant here and the other there, but both generally present and almost inextricably combined in the evolution of any particular group. The orienting element was found rather surely to be adaptation, and not such proposed alternatives as innate life tendency or progression toward a destined goal according to plan. Since it has been found that the materials for change are largely random, the question naturally arises as to how evolutionary change can be so extensively nonrandom, how adaptation can orient it.

The mechanism of adaptation is natural selection. The idea of natural selection is very simple, even though its operation is highly complex and may be extremely subtle. Natural selection has this basis: in every population some individuals have more offspring than others. This obvious fact automatically accounts for the possibility of evolutionary change.

It has been seen that individuals in any group differ in genetic make-up, hence pass on different heredity to their offspring, and also that mutations occur in a scattered way as such a group reproduces. It may happen, and is indeed the usual thing over small numbers of generations, that the new generation, in spite of its differences among individuals, has about the same average genetic constitution as the parent generation and about the same incidence of mutations, so that no clearly evident change occurs from one generation to the next. It is, however, extremely unlikely that the new generation will have exactly the same genetic make-up as the parental generation. Some individuals do have more offspring than others and their particular genetic characters, which differ to some degree from those of other individuals, will be more frequent in the new generation. The difference may be imperceptibly slight or it may be quite marked, even in production of a single generation.

Even a very slight change will produce evident, eventually large effects if it is cumulative from one generation to the next. This results if there is some constant factor such that, on an average, the individuals that do have more offspring in each generation are those tending toward the same hereditary type; in other words, if they are somehow selected for characteristics that have a hereditary basis. In nature the individuals that tend to have more offspring are, as a rule and no matter how slight the difference, either those best integrated with their environment (including the association with their own species) and most successful in it or those best able to begin to exploit an opportunity not available or less so to their neighbors. Thus natural selection usually operates in favor of maintained or increased adaptation to a given way of life, organism-environment integration, or alternatively of such change as will bring about adaptation to another, accessible way of life. Natural selection thus orients evolutionary change in the direction of one or another of these sorts of adaptation. We have seen, in fact, that these are the usual directions of orientation in evolution, to the extent that such orientation is effective.

Natural selection as it was understood in Darwinian days emphasized "the struggle for existence" and "the survival of the fittest." These concepts had ethical, ideological, and political repercussions which were and continue to be, in some cases, unfortunate, to say the least. Even modern students of evolution have not always fully corrected the misconceptions arising from these slogans. It should now be clear that the process does not depend on "existence" or "survival," certainly not as this applies to individuals and not even in any intensive or explanatory way as it applies to populations or species. It depends on differential reproduction, which is a different matter. It does not favor "the fittest," flatly and just so, unless you care to circle around and define "fittest" as those that do have most offspring. It does favor those that have more offspring. This usually means those best adapted to the conditions in which they find themselves or those best able to meet opportunity or necessity for adaptation to other existing conditions, which may or may not mean that they are "fittest," according to understanding of that word. Moreover the correlation between those having more offspring, and therefore really favored by natural selection, and those best adapted or best adapting to change is neither perfect nor invariable; it is only approximate and usual.

It is, however, the word "struggle" that has led to most serious misunderstanding of the process of natural selection, along with a host of related phrases and ideas, "nature red in fang and claw," "class struggle" as a natural and desirable element in societal evolution, and all the rest. "Struggle" inevitably carries the connotation of direct and conscious combat. Such combat does occur in nature, to be sure, and it may have some connection with differential reproduction.

A puma and a deer may struggle, one to kill and the other to avoid being killed. If the puma wins, it eats and presumably may thereby be helped to produce offspring, while the deer dies and will never reproduce again. Two stags may struggle in rivalry for does and the successful combatant may then reproduce while the loser does not. Even such actual struggles may have only slight effects on reproduction, although they will, on an average, tend to exercise some selective influence. The deer killed by the puma may be too old to reproduce; if the puma does not get the deer, it will eat something else; the losing stag finds other females, or a third enjoys the does while the combat rages between these two.

To generalize from such incidents that natural selection is over-all and even in a figurative sense the outcome of struggle is quite unjustified under the modern understanding of the process. Struggle is sometimes involved, but it usually is not, and when it is, it may even work against rather than toward natural selection. Advantage in differential reproduction is usually a peaceful process in which the concept of struggle is really irrelevant. It more often involves such things as better integration into the ecological situation, maintenance of a balance of nature, more efficient utilization of available food, better care of the young, elimination of intragroup discords (struggles) that might hamper reproduction, exploitation of environmental possibilities that are not the objects of competition[4] or are less effectively exploited by others.

It is to be emphasized that the group of its own kind among which an animal lives is also a part of its environment,

4. The word "competition," used in discussion here and previously, may also carry anthropomorphic undertones and then be subject to some of these same objections. It may, however, and in this connection it must, be understood without necessary implication of active competitive behavior. Competition in evolution often or usually is entirely passive; it can occur without the competing forms ever coming into sight or contact.

but a special part. There is an intraspecific selection, based on integration and association within the group, as well as extraspecific selection, based on adaptive relationship to the environment as a whole. (The same sort of distinction was made in discussing specialization and competition in relationship to extinction.) Intragroup selection may involve actual struggle, as in the case of the stags fighting for a doe. It may then be deleterious as regards extragroup adaptation and involve selection opposed to extragroup selection. If such is the case, the result, as Haldane has emphasized, may be deleterious for the species as a whole, although even here we may remark that intra- and extragroup struggle commonly produce selection in the same direction. It is to be added that in intragroup selection, also, struggle is not necessarily or even usually of the essence. Precisely the opposite, selection in favor of harmonious or cooperative group association, is certainly common.

It was a crude concept of natural selection to think of it simply as something imposed on the species from the outside. It is not, as in the metaphor sometimes used with reference to Darwinian selection, a sieve through which organisms are sifted, some variations passing (surviving) and some being held back (dying). It is rather a process intricately woven into the whole life of the group, equally present in the life and death of the individuals, in the associative relationships of the population, and in their extraspecific adaptations.

A criticism formerly and even now occasionally leveled against Darwinian natural selection, but no longer valid against the modern conception of this process, was that it seemed to be a purely negative factor in evolution and one that could not account for positive aspects in the orientation of evolution. Some role for selection was admitted, as it must be, but it was maintained that this role was merely the elimination of some individuals or types of organisms: the "unfit." Such negative action seemed to have no particular bearing on the origin of new types or on the maintenance of positive adaptive change in evolutionary trends. This negative aspect does exist, but it is now evident that selection also has a positive and creative role and that it is indeed the decisive, the orienting, process in continuing adaptation. Part of the difficulty lay again in thinking in terms of struggle and survival, the death of one animal and the triumph of its enemy or rival. The concept of evolution as change in proportions, combinations, and nature of genetic factors in populations is of more recent development and this entails consequences that were not understood by the critics of natural selection as the guiding force in evolution.

Selection is not primarily a process of elimination. It is a process of differential reproduction and this involves complex and delicate interplay with those genetic factors in populations that are the substantial basis of evolutionary continuity and change. In terms of single, arising mutations, those that are unfavorable will be eliminated by selection, as far as its force is effective. (The theory does not demand and the facts do not indicate that selection is always effective, or that at its most effective it can eliminate all unfavorable mutations immediately.) Those that are favorable will, however, tend under the influence of selection to spread through the population increasingly in successive generations. This, in itself, is a positive evolutionary change which is due to natural selection.

Further, the characteristics of an organism as an integrated whole depend not on the action of one gene or another but on the whole interacting set of genes. Some combinations produce poorly integrated organisms, others well-integrated ones. There are, moreover, myriads of different possible combinations even within one population that are all capable of producing workable integration but that produce each a different variant in type. As has been mentioned, the possible number of such combinations vastly exceeds the whole number of individual organisms that can ever exist within a given population or species. Only a fraction of the possible combinations is actually realized in a concrete, existing organism. The chances of such realization for any particular combinations depend on the frequencies in the parental population of the genes involved in the combination. These frequencies, in turn, are to considerable extent, if not absolutely, determined by the action of natural selection. Selection thus plays an essential part in determining what combinations of genes will be incorporated in individual organisms and so in the actual origin of new variant sorts of organisms. In this role, selection may surely be said to be a creative factor in evolution. Still further, once such favorable or adaptive combinations have arisen, selection tends to hold them together, to keep them from being shuffled apart again in the random processes of heredity in reproduction, and tends to promote their spread through the population.[5]

The way in which selection and the other factors and forces of evolution interact within a population was worked out in a brilliant, now classical series of studies by R. A. Fisher, J. B. S. Haldane, Sewall Wright, and others.[6] It was these studies, as much as anything else, that made possible the synthesis of generations of observations and experiments in a wide variety of fields into a coherent and comprehensive

5. This creative aspect of selection has been discussed more fully in G. G. Simpson "The Problem of Plan and Purpose in Nature" (*Sci. Monthly, 64* [1947], 481–495), revised and extended in *This View of Life* (New York, Harcourt, Brace and World, 1964).

6. R. A. Fisher, *The Genetical Theory of Natural Selection* (Oxford, Clarendon Press, 1930); J. B. S. Haldane, *The Causes of Evolution* (New York and London, Harper, 1932); S. Wright, "Statistical Genetics and Evolution" (*Bull. Amer. Math. Soc., 48* [1942], 223–246). These authors and others have written numerous other studies on aspects of this subject. The cited paper by Wright is only one in a long sequence of short contributions made over a long period of years which unfortunately have not been gathered into a comprehensive publication by their author.

modern theory of evolution. For the reader who may have browsed widely in the literature of evolution and who may have become bewildered by continuing divergences of opinion, here is a touchstone: I think it fair to say that no discussion of evolutionary theory need now be taken seriously if it does not reflect knowledge of these studies and does not take them strongly into account.

The evolutionary materials involved in this complex process are the genetical systems existing in the population and the mutations arising in these. The interacting forces producing evolutionary change from these materials are their shuffling in the process of reproduction, the incidence of mutations (their nature and rate), and natural selection.

The genetical systems existing in a population determine (in further interaction with the environment in each individual case) the nature of the organisms comprising that population. The variety of these systems—and they always are more or less varied—determines the variability of the population. This has extreme importance for evolution, because it is directly on or through this variability that natural selection operates. Limited variability offers correspondingly limited chance of rapid change or of local adaptation to particular conditions, but on the other hand in a population that is well adapted it holds a larger proportion of the population at an optimum, the best adaptive type under existing conditions. Wide variability offers the possibility of rapid (but not indefinitely continued) evolutionary change, quick adjustment to environmental changes, and local adaptation to special conditions within the wider range of the whole population. But it means that some proportion, perhaps a considerable proportion, of the population will deviate markedly from the optimum condition, will not be as well adapted as they might be. As in so many evolutionary phenomena there is here a complex interplay, a balancing of opposed advantages and of other factors.

Another important fact to be mentioned only in passing in this necessarily brief review of a very complicated subject is the existence in populations of hidden variability, which can be evoked under the influence of selection or in other ways. This involves the usual presence in populations of genes (especially the recessives of the geneticists) whose action is prevented or overlain in the presence of other genes (especially the dominant alleles of the geneticists) which may be more common in the population. If the general proportion of genes with hidden effects rises in the population, they will more often occur in pairs in the same individual (which is then recessive homozygous in genetical terms) rather than in combination with their concealing equivalent genes (in heterozygous combination), and their effect will then appear. The always great potential for new recombinations of genetic factors is also and usually to even greater extent, a source of hidden or previously unmanifested variability, especially as the expression of individual genes may be different in different genetic contexts or combinations.

In a population indefinitely large, breeding wholly at random, not affected by selection, the proportions of the various existing sorts of genes and chromosomes tend to maintain definite fixed ratios. While this equilibrium lasts, change in genetic ratios and consequently evolutionary change are slight or nil. The population will still vary, but the variability will be constant in nature and extent. This condition does not occur and evolutionary change does ensue because populations are of limited size, because they do not breed entirely at random, because new sorts of genes and chromosomes do arise, and because selection does act on the population.

The outcome of the shuffling process and the effectiveness, absolute and relative, of mutation and selection are largely dependent on the size and breeding structure of the population. Almost any variation in genetic systems and almost any mutation will be affected to some extent by selection, even though the effect may be slight. It is rare for variations and mutations to have no selective advantage or disadvantage whatsoever. Whatever the influence of selection may be, there is always some chance that random changes will become established in a certain proportion of the population or even that they will spread to the whole population in succeeding generations. This is one reason why evolution is not completely adaptive and does show random influences. Other reasons are that selection can act only on variations that do occur, which are not necessarily or usually the best possible from the point of view of adaptation, and that a given mutation, for instance, commonly has multiple effects some of which are adaptive and others nonadaptive or inadaptive.

Selection almost always has some effect on the fixation or elimination of favorable or unfavorable changes. The chances of fixation in a population of given size and structure are proportional to intensity of selection for or against, although the chances are very rarely quite 100 per cent or 0. For a given intensity of selection, these chances tend to vary with size of the breeding population. The exact relationships are, again, rather complicated, but the general tendency is for selection to be more effective the larger the population. In large interbreeding populations the chances of spread of a variation purely at random become very slight: favorable selection, even if very slight in degree, will usually ensure such spread and unfavorable selection will usually prevent it. Evolution in large populations is dominated by selection, tends to be closely proportional to the intensity of selection, and tends to have only few and extremely slow changes that are not purely and directly adaptive. It is just such groups that do show, in their evolutionary records, clear and long-continued trends adaptive in control. It is also such groups, with large and widespread populations, that tend to leave more abundant and more continuous fossil records, which doubtless helps to account for the impression of some paleontologists that evolution usually or always follows well-defined trends.

On the other hand, in small breeding populations the effectiveness of selection may be reduced. The chances of merely random change are correspondingly increased. In very small populations, of only a few dozens or perhaps at most hundreds of individuals, evolutionary change may be mainly at random and the influence of selection may be almost negligible unless the intensity of selection is unusually great. Such random evolution is almost always inadaptive and its usual outcome is extinction.

Continuous and evenly distributed interbreeding within a whole population (say the whole of a given species) is not the rule in nature or in human society and, indeed, is perhaps never fully exemplified. In the usual situation, there are local groups that habitually interbreed, with greater or less inbreeding of their various family strains. (In man, these groups are defined not only by geographic proximity but also by social and occupational status, intellectual level, and other societal factors such as religion.) Between such groups there is a certain but limited amount of crossbreeding, with consequent flow of genetic factors between groups through the generations. The ideal situation for maintenance of continuously good adaptation, including rapid evolution when adaptive change is possible or necessary, seems to involve a relatively large total population divided into a large number of moderate-sized local breeding groups, with some continual gene interchange (crossbreeding) between adjacent groups.

With this understanding of the evolutionary process as it really works in populations of organisms and on the factors that determine their heredity, adaptation ceases to be a miracle, or even a serious problem. It is adaptation that gives an appearance of purposefulness in evolution and in its results in the present world of life. Its explanation was the main stumbling block in acceptance of evolution as a fact, and later in attempts to explain the course of evolution by one theory or another. Failures of earlier naturalistic attempts to explain this apparent purposefulness were responsible, in large part, for the conclusions of some students that this betokened purpose, in fact, and of certain among these that the purpose betokened a Purposer. It would be brash, indeed, to claim complete understanding of this extraordinarily intricate process, but it does seem that the problem is now essentially solved and that the mechanism of adaptation is known. It turns out to be basically naturalistic, with no sign of purpose as a working variable in life history, and with any possible Purposer pushed back to the incomprehensible position of First Cause.

A constant stumbling block in the way of attempts to understand evolution has been that its processes must explain not only adaptation but also absence of adaptation, the existence and persistence of apparently random as well as of clearly oriented features in evolution. This was, and remains, an unanswered argument against theories demanding the reality of purpose or the existence of a goal in evolution. It equally renders untenable all the other theories that attempted to explain evolution by the dominant or exclusive action of one single principle or another, such as the Neo-Darwinian insistence on natural selection as essentially the whole story. Modern understanding of evolution is not as simple as were these various theories, but their simplicity was factitious. They were bound to be wrong in seeking a simple explanation for something that is, in its nature and its phenomena, so far from simple.

The presence, often simultaneously, of both adaptive and nonadaptive, both apparently purposeful and apparently purposeless, both oriented and random features in evolution has now been sufficiently explained. Nonadaptive and random changes have another possible role in evolution that is important and that has so far been suggested only in passing. They have a bearing on changes in broad types of organization, the appearance of new phyla, classes, or other major groups in the course of the history of life. The process by which such radical events occur in evolution was the subject of one of the most serious disputes among qualified professional students of evolution. The question is whether such major events take place instantaneously, by some process essentially unlike those involved in lesser or more gradual evolutionary change, or whether all of evolution, including these major changes, is explained by the same principles and processes throughout, their results being greater or less according to the time involved, the relative intensity of selection, and other material variables in any given situation.[7]

Possibility for such dispute exists because transitions between major grades of organization are seldom well recorded by fossils. There is in this respect a tendency toward systematic deficiency in the record of the history of life. It is thus possible to claim that such transitions are not recorded because they did not exist, that the changes were not by transition but by sudden leaps in evolution. There has been much diversity of opinion as to just how such leaps might have happened. Beurlen, for instance, ascribed them vaguely

7. The former opinion was adopted by Schindewolf in his work previously cited, and was also strongly supported by K. Beurlen, *Die stammesgeschichtlichen Grundlagen der Abstammungslehre* (Jena, Gustav Fischer, 1937); and R. Goldschmidt, *The Material Basis of Evolution* (New Haven, Yale University Press, 1940). The subject was particularly well reviewed and the arguments of Schindewolf, Beurlen, Goldschmidt, and others of their school strongly and I think conclusively refuted in Rensch, *op. cit.* In fact there is now almost no support for that view except by a few philosophers not sufficiently acquainted with scientific data on evolution.

and vitalistically to an inner urge or will on the part of the organisms concerned. Goldschmidt ascribed them to a sudden over-all remodeling of the genetic system, a "systemic mutation" different in kind from the well-known gene and chromosome mutations of more orthodox genetics. Schindewolf thought that they are mutations, apparently of the usual sort, but large mutations that markedly change the course of individual development from its early stages and thus produce radically new adult forms.

It is impossible in a brief space, and it is unnecessary now, to discuss the pros and cons of this argument in detail. Enough has, indeed, already been said to throw some of the more extreme views out of court without further hearing. "A will toward individualization and independence" is a resplendent phrase but one essentially meaningless in the face of actual evolutionary processes as these are now known. "Systemic mutations," which have never been observed and the supposed nature of which has not been concretely described, need not be taken seriously if, as is the case, the phenomena that they were postulated to explain can be explained in terms of known processes and forces.[8] If only ordinary, but "large," mutations are supposed to be involved, then we are back within the framework of the modern consensus, although of course we would still like to know, and will still be constrained to discuss, the relative parts played in evolution by large and small mutations.

Transitional types are not invariably lacking in the record.

8. In some of his latest work Goldschmidt implicitly retreated from his position by suggesting that his theory differed from that of almost all his genetical colleagues as regards only the size of the mutations involved, not their nature. If this emendation be accepted, the difference of opinion, as with Schindewolf's views in contrast with those of most other paleontologists, no longer has any fundamental importance for evolutionary theory, although most of the many and strong objections to his views still stand.

A multitude of them are known between species, many between genera, a few between classes, but none, it is true, between phyla. Most of the phyla appear toward the beginning of the Paleozoic, as discussed in Part I, and the paucity of record of prior ancestral types is as hard or as easy to explain whether we suppose that they arose instantaneously or gradually. The record is obviously a sampling only and full of gaps. We would suppose that if all changes were by slow transition we still would find only a small proportion of the transitional types and might find none between the phyla, few in number and with their special conditions of early preservation. On the other hand, if major changes were always instantaneous, obviously we should find no transitional types—and we do find many of them. If we did not happen to have found such types between fishes and amphibians, amphibians and reptiles, reptiles and birds on one hand and mammals on the other, or even between eohippus and the horse, these particular changes would surely be considered instantaneous by students who incline to that view. As H. E. Wood has remarked, the argument from absence of transitional types boils down to the striking fact that such types are always lacking unless they have been found.

As far as analysis has been carried on living forms, distinct populations, with their separate characteristics, do not customarily differ in presence or absence of single mutations but in having different, integrated genetic systems, which may involve differences in dozens, hundreds, or thousands of individual genes. On this point Goldschmidt's analysis seems sound and extremely suggestive, although his conclusion that therefore the difference arises by "systemic mutation" is a non sequitur. A mutation may produce discrete differences and to this extent its appearance is an instantaneous and discontinuous evolutionary event, whether its effects be small or large. But it is populations, not individuals, that evolve. For a given mutation, regardless of its "size," to become involved in the origin of a new and especially of a highly distinctive group of animals it must spread through a population and while doing so and thereafter it must become integrated in a new sort of genetic system.[9] It is very nearly impossible to imagine these processes occurring except by transition over a long sequence of generations, and certainly no conclusive, or even really suggestive, opposite example is provided by the paleontological record.

It is evident that these processes, which are normal in evolution, could possibly occur with large mutations as well as small. They are, however, far more likely to occur with small mutations than with large. The chance that a mutation will be favored by selection and the chance that it will or can be integrated into a genetic system as a whole are inversely related to the effect of the mutation on the organism. If this effect is really radical, comparable, say, to the difference between one family and another (*a fortiori,* to that between higher categories) in the recent fauna, the chances that the mutation will really take, so to speak, and lead to an evolutionary progression are so small as to be almost negligible. On the other hand, the cumulative effect of mutations so small as to have almost imperceptible effects, each spread in the population and each successively integrated into the genetic system under the influence of selection, is more than adequate to account for transitions from one structural grade to another in the time that the record shows was expended, or was available, in such cases.

It is thus likely, to say the least, that major as well as minor changes in evolution have occurred gradually and that the same forces are at work in each case. Nevertheless there is a

9. This special aspect of evolution was treated in C. D. Darlington, *The Evolution of Genetic Systems* (Cambridge, England, Cambridge University Press, 1939). Among more recent studies is V. Grant, *The Architecture of the Germplasm* (New York, Wiley, 1964).

difference and many of the major changes cannot be considered as simply caused by longer continuation of the more usual sorts of minor changes. For one thing, there is excellent evidence that evolution involving major changes often occurs with unusual rapidity, although, as we have seen, there is no good evidence that it ever occurs instantaneously. The rate of evolution of the insectivore forelimb into the bat wing, to give just one striking example, must have been many times more rapid than any evolution of the bat wing after it had arisen. The whole record attests that the origin of a distinctly new adaptive type normally occurs at a much higher rate than subsequent progressive adaptation and diversification within that type. The rapidity of such shifts from one adaptive level or equilibrium to another has suggested the name "quantum evolution," under which I have elsewhere discussed this phenomenon at greater length.[10]

Another peculiarity of such evolutionary events is that they always represent distinct changes in the direction of evolution. Such changes may occur even though the trend is continuously adaptive and its control is by natural selection throughout. Change from browsing to grazing in horses is a clear and fully documented example. The same is probably true of changes more radical in character and usually less well documented. The now fairly well-known fish-amphibian transition, for instance, has no probably nonadaptive features as regards the essential changes involved. Yet we sense in many such changes that here the random element in evolution has been involved. The change-over from reptile to mammal, for instance, involved a long and adaptive trend in the reptiles, but at the last, in the switch from one type of

10. Originally in *Tempo and Mode in Evolution* (New York, Columbia University Press, 1944) and then in *The Major Features of Evolution*, previously cited. The term "quantum" in this connection has been subjected to some criticism and indeed it is not very satisfactory, but no one has yet suggested a better.

ear to the other, for instance, there seems to be a sort of non sequitur or experimentalism. The broadly opportunistic pattern of nature also suggests something of the same thing.

In such cases and in other instances of rather rapid change, great or small, in adaptive type, it is possible, at least, that random preadaptation has occurred. The possibility of this process has long been recognized and many geneticists formerly assigned to it a wide, even an all-embracing importance in evolution, which was certainly much overemphasized. By "preadaptation" is generally meant the random origin, by mutation, of characteristics nonadaptive or inadaptive for the ancestral way of life, but adaptive for some other way of life which happens to be available.[11] As a rule the spread, integration, and concomitant utilization of such a change could occur only under the influence of selection, and then this is only a special case of adaptation—adaptation changing its direction—and to call it preadaptation may merely be juggling with words and not establishing a real distinction. It is, however, possible for mutations occasionally to become established without benefit of selection, or even in the face of adverse selection, especially in very small populations. Almost always this would lead to extinction, for the group could not survive long if definitely and continuously disadvantageous characters became established. In the odd case permitted by the qualification "almost," however, the characters disadvantageous in a current way of life might become advantageous when a change in way of life was possible and occurred. Preadaptation in that sense, although extremely rare in evolution (no probable example is really known) might provide a mechanism for sudden and erratic changes in adaptive type. The importance of such an event

11. Interesting discussion and exemplification of this phenomenon, written when it was being given undue importance, will be found in L. Cuénot, *L'Adaptation* (Paris, G. Doin, 1925). More moderate, more maturely considered discussion was given in Cuénot and Tétry (1951, previously cited).

could be great, even though its occurrence were markedly exceptional. Rise of radically new types of organisms *is* exceptional.

At the other end of the scale of evolutionary phenomena are the features of deployment of populations, of their adaptations to merely local variations in conditions, and of their splitting into two or more discrete populations when interbreeding ceases between subdivisions of them. Under the name of "speciation," these processes have been more intensively studied than any other aspects of the great subject of evolution, and indeed some students seem to consider "speciation" as practically synonymous with "evolution." This orientation of study has been inevitable and, in large degree, desirable. The fundamental proposition of the doctrine of special creation was the immutability of species, so that the first essential in establishing the truth of evolution was to demonstrate that one species may give rise to another. Hence the title of Darwin's great work on evolution in general, *The Origin of Species.* Variations, heredity, and processes of differentiation within species can be studied experimentally, whereas the larger features of evolution cannot. In practice, field naturalists and zoologists in museums are mainly occupied in distinguishing and examining species and their subdivisions, so much more numerous than the broader groups and so much more difficult to recognize.[12] More broadly, in dealing with recent animals, this is the best place to attack problems of evolution, where they begin, within populations and in their readily observed characteristics and activities. Here the basic theories of population genetics, partially summarized above, were worked out.

Aspects of speciation that bear essentially on our present theme, that of the general features of the history of life and the meaning of evolution, have already been discussed by

12. The best discussion of speciation from this point of view, together with copious citations from the enormous literature on speciation in general, is E. Mayr's previously cited book of 1963.

implication if not explicitly. The subject has other aspects that are both fascinating and important, particularly the question of isolating mechanisms, the evolutionary devices for preventing interbreeding between closely related and formerly united populations. These aspects are not, however, absolutely necessary to the development of the broader subject of this book and adequate treatment of them would itself extend to book length.

One thing more may, however, be said about speciation in relationship to the processes of evolution here more extensively or explicitly treated. In Huxley's barrel of life speciation is the final pouring in of water, the filling up of the interstices in the long process of packing every available environment with as much life as it can support. To avoid the danger of analogy, it must, however, be emphasized that speciation was by no means the last thing to occur in evolution. It was equally the first, for it goes on all the time, along with any other evolutionary process that may be under way. It is the basic structure of the web of life, the ever-present detail of the fabric of evolution. It is the source and the fundamental process of diversification in the forms of organisms. Its pattern is that of branching, the separation of one group of organisms from another and the distribution between them of different portions of the stock of hereditary variation.

In the actual process of descent and change in the long history of life, speciation is one of the essential patterns. Along with this, and always embodying this within them, are two other patterns of constant occurrence and major importance. One of these is the pattern of trend, progressive, oriented change under the control of adaptation. The other is the more rapid and sporadic, recurrent rather than continuous, pattern of change in adaptive type, adoption of a new and distinct way of life. In all these interwoven patterns the same evolutionary forces are at work, but they vary endlessly in intensity, in combination, and in result.

XV. THE CONCEPT OF PROGRESS IN EVOLUTION

It is impossible to think in terms of history without thinking of progress. With reference to what is at the moment, or to structure and relationships outside a framework of time, the concept of progress is irrelevant or at least unobtrusive. Classic chemistry and physics were sciences without time and the idea of progress simply did not arise in connection with, say, the union of sodium and chlorine atoms in a salt molecule. Even in these relatively nonhistorical sciences the newer concepts of the space-time continuum and the transmutation of elements through the periodic system introduce time, demand a historical approach, and involve ideas of progression or retrogression.

Life is so obviously a process in time and not merely a static condition of being that this study has always to some degree involved historical concepts. Development and progression are so plainly evident in animate nature that these features deeply impressed biologists long before the grand fact of the evolution that produced them was understood. The idea of biological progress is as old as the science of biology and it was already deeply imbedded in pre-evolutionary science. Although its actual historicity and its real relationship to the flow of time were scarcely glimpsed, this concept of a progression of life from lower to higher was fundamental both in primitive theology (such as the Semitic creation myths) and in primitive science (such as that of Aristotle), and it was taken over, more or less as a matter of course, in later pre-evolutionary biology that still stemmed, in the main, from these two sources. Evolution, revealing the development of life as an actually and materially historical process, gave meaning to these older observations and to the almost intuitive concept of progression if not, fully, of progress. The first truly general theory of evolution, that of Lamarck, had as its central feature the very ancient and previously nonevolutionary idea of a sequence of life forms from less to more perfect.

Examination of the actual record of life and of the evolutionary processes as these are now known raises such serious doubts regarding the oversimple and metaphysical concept of a pervasive perfection principle that we must reject it altogether. Yet there is, obviously, progression in the history of life, and if we are to find therein a meaning we are required to consider whether this involves anything that we can agree to call "progress," and if so, its nature and extent.

It is a childish idea—but one deeply ingrained in our thinking, especially on political and social subjects[1]—that change *is* progress. Progression merely in the sense of succession occurs in all things, but one must be hopelessly romantic or unrealistically optimistic to think that its trend is necessarily for the good. We must define progress not merely as movement but as movement in a direction from (in some sense) worse to better, lower to higher, or imperfect to more nearly perfect. A description of what has occurred in the course of evolution will not in itself lead us to the identification of progress unless we decide beforehand that progress *must* be inherent in these changes. In sober enquiry,

1. The concept of evolutionary organic progress has been strongly influenced and reinforced by the even older idea of social progress, which became an essential part in the climate of opinion of the eighteenth and nineteenth centuries. It has been challenged in our own century, but still is the prevailing view in industrialized countries at least, both capitalist and communist. See, for example, B. Mazlish, "The Idea of Progress" (*Daedalus*, *92* [1963], 447–461).

we have no real reason to assume, without other standards, that evolution, over-all or in any particular case, has been either for better or for worse. Progress can be identified and studied in the history of life only if we first postulate a criterion of progress or can find such a criterion in that history itself.

The criterion natural to human nature is to identify progress as increasing approximation to man and to what man holds good. The criterion is valid and necessary as regards human history, although it carries the still larger obligation of making a defensible and responsible choice among the many and often conflicting things that men have held to be good. The criterion is also perfectly valid in application to evolution in general, provided we know what we are doing. Approximation to human status is a reasonable *human* criterion of progress, just as approximation to avian status would be a valid avian criterion or to protozoan status a valid protozoan criterion. It is merely stupid for a man to apologize for being a man or to feel, as with a sense of original sin, that an anthropocentric viewpoint in science or in other fields of thought is automatically wrong. It is, however, even more stupid, and even more common among mankind, to assume that this is the *only* criterion of progress and that it has a *general* validity in evolution and not merely a validity relative to one only among a multitude of possible points of reference.

On the other hand we may find, and we will as the discussion proceeds, that criteria *not* selected with man as the point of reference still indicate that man stands high on various scales of evolutionary progress. It would then be foolish to cry "Anthropomorphism!" After all, it may be a fact that man does stand high or highest with respect to various sorts of progress in the history of life. To discount such a conclusion in advance, simply because we are ourselves involved, is certainly as anthropocentric and as unobjective as it would be to accept it simply because it is ego-satisfying. We should now look with amazement or with condescension on the once more general attitude among scientists that a sense of values, and especially one that values our own species, is unscientific.

As a start in the enquiry, it is quickly evident that there is no criterion of progress by which progress can be considered a *universal* phenomenon of evolution. There are the cases in which change, and therefore any possible sort of progress, has been arrested except for minor and local, not progressive, fluctuations. There are also few possible definitions of "progress"—I think only one, and that one not really acceptable—under which the term could be applied both to the rise of the marvelously intricate organism of a typical crustacean, such as a crab, and to the change of this organism, in connection with parasitism, to an almost formless mass of absorptive and reproductive cells. Whatever criterion you choose to adopt, you are sure to find that by it the history of life provides examples not only of progress but also of retrogression or degeneration. Progress, then, is certainly not a basic property of life common to all its manifestations. This casts further doubt (at least) on the finalist thesis, still more on the concept of a perfecting principle, but it certainly does not justify a conclusion that progress is absent in evolution. In a material world the very idea of progress implies the possibility of its opposite. To find that progress is universal would certainly be far more surprising than to find that it is only occasional.

In considering the record of life, we sought progressive changes that involve life as a whole, and not only the evolution of particular groups within the total process. Only one was found: a tendency for life to expand, to fill in all the available spaces in the livable environments, including those created by the process of that expansion itself. This is one possible sort of progress. Accepting it as such, it is the only one that the evidence warrants considering general in the course of evolution. It has been seen that even this, although general, is not invariable. The expansion of life has not been constant and there have even been points where it lost ground temporarily, at least. The general expansion may be considered in terms of the number of individual organisms, of the total bulk of living tissue, or of the gross turnover, metabolism, of substance and energy.[2] It involves all three, and increase in any one is an aspect of progress in this broadest sense.

This general expansion is only imperfectly helpful in providing a criterion of progress applicable to a particular case. Any group that has persisted and that does make up part of the sum total of existing life must be granted a share in this progress and from this point of view protozoan and man stand on a level. There are, however, different degrees of contribution to the expansion. The little sphenodon on its islands off New Zealand is certainly not contributing very much to the filling of the earth with life. Man is making a large contribution, not only in the persons of his bulky millions but also—almost uniquely among all organisms—in his vast swarm of domestic plants and animals. A criterion is also provided for evolutionary movement within any given group. As one group replaces another the net total of life may

2. This last aspect, particularly, is discussed in the paper by Lotka cited earlier. Other tendencies that have sometimes been considered universal in evolution include: increase in complexity (most writers on evolution, see hereafter); decrease in entropy, for example B. C. Patten, "An Introduction to the Cybernetics of the Ecosystem: the Trophic-dynamic Aspect," (*Ecology, 40* [1959], 221–231); minimization of effort, for example G. K. Zipf, *Human Behavior and the Principle of Least Effort* (Cambridge, Addison-Wesley, 1949); and increase in homeostasis, for example, A. E. Emerson, "Dynamic Homeostasis: a Unifying Principle in Organic, Social, and Ethical Evolution" (*Scientific Monthly, 78* [1954], 67–85). All of these have occurred at times in the course of evolution, but none is invariable or can be considered a valid generality about evolution.

or may not be changed, but by this particular criterion alone, within the history of one group, its increase in variety and abundance may be considered ipso facto progressive and its decline retrogressive. In this sense extinction is not merely the end but also the very antithesis of progress—but ultimate termination (the inevitable fate of all life) is no sign that progress was earlier absent in the rise and history of the group. As regards direction and intensity of expansion at any one time, man is right now the most rapidly progressing organism in the world. The actual bulk of material incorporated in *Homo sapiens* seems now quite clearly—and from other points of view, most unfortunately—to be increasing more rapidly than in any other species.

In the search for, as nearly as may be, objective criteria of progress applicable to a particular case and yet widely valid, such considerations lead to the criterion of dominance, which was at one time stressed by the leading student of the subject, J. S. Huxley.[3] We have seen how throughout the history of life each group has tended to expand and to have one or more periods when it was particularly abundant and varied. We have seen, too, how at any given time certain groups tended to be much more varied and abundant than others, in other words to dominate the life scene, and that there has been a succession in these dominant groups. It is this succession that provides a criterion, though plainly not the only one, of evolutionary progress. Thus among the aquatic vertebrates, it is fully justified, as long as we keep in mind the particular *kind* of progress that we mean, to say that successive dominance of Agnatha, Placodermi, and Osteichthyes represents progress and that Osteichthyes are

3. In *Evolution, the Modern Synthesis* (first edition, 1942; second edition, reprinted main text but with a lengthy, modernizing new introduction, 1963, previously cited). I am greatly indebted to that and many other works by Sir Julian. He later (that is, after 1942) developed other ideas, or aspects, of progress, to be noted hereafter.

the highest, Agnatha the lowest, group among the three. It is worthy of note, in passing, that the Chondrichthyes are generally admitted in this sequence as higher than Placodermi but lower than Osteichthyes, yet the Chondrichthyes by *this* criterion are really neither lower nor higher than the Osteichthyes. The same sort of sequence applies in the successive dominance of Amphibia, Reptilia, and Mammalia.

We do not, however, find successive dominance among, say, Osteichthyes, Aves, and Mammalia. All three are dominant at the same time, during the Cenozoic and down to now. Taking the animal kingdom as whole, it is clearly necessary to add insects, molluscs, and also the "lowly" Protozoa as groups now dominant. If one group had to be picked as most dominant now, it would have to be the insects, but the fact is that all these groups are fully dominant, each in a different sphere. The conclusion is emphasized by cases like that of the Osteichthyes and the Amphibia—the Osteichthyes were dominant later. If this criterion were really given general validity and were objectively applied, it would be necessary to conclude that bony fishes are higher forms of life and represent a further degree of progress than amphibians, or than reptiles, for that matter.

It is true, as Huxley says, that "biologists are in substantial agreement as to what were and what were not dominant groups" and that they usually arrange these as if they constituted a single succession symbolized by the stereotyped "Age of Invertebrates," "Age of Fishes," "Age of Amphibians," "Age of Reptiles," "Age of Mammals," and "Age of Man." The fact is that such a sequence does not follow a single or a wholly objective criterion. It is not based solely, or even in the main, on the objective facts of dominance in the history of life. Two other criteria have been sneaked in: that of ancestral and descendant relationship, which is an objective and general criterion but an entirely different matter from dominance, and that of approximation to man, which also has nothing to do with the facts of dominance and which, although perfectly all right as long as we know what we are doing and do not think we are following a general and objective criterion, is specific to a single point of reference and subjective to that point, i.e., man.

The criterion of dominance is not invalidated by the fact that it does not yield a single sequence for progress in evolution. That there is or should be a single sequence was a fallacy of the pre-evolutionary perfecting principle and "ladder of nature" ideas, and this can be a required condition of progress only when progress is defined with reference to a point and not as a general principle. Of course even as a general principle it *might* have been found that progress was a one-line affair, and should be if there were anything in most vitalist or finalist contentions, but as a matter of fact this does not turn out to be the case with the criterion of dominance or, as will be found, with any other criterion *except* that of approximation to some specific type such as man.

The various different lines of progress that are involved in successive dominance are defined by adaptive types or corresponding ways of life. Thus there is broadly one for aquatic vertebrates and broadly another, quite distinct from this, for terrestrial vertebrates, with bony fishes at the top in the former and mammals simultaneously at the top in the latter. Within each of these groups there are other, more closely circumscribed ways of life and corresponding lines of dominance and of progress (in this sense) for each of these. Among the bony fishes, there are separate lines for marine and freshwater fishes, for shallow and deep sea fishes, and so on.

In relationship to man, not taking him as point of reference but adhering to strict dominance as objective criterion, we find that he is a member of a progressive group, and generally of the most progressive, in each of the various dominance sequences in which he can properly be placed. (His dominance cannot be compared with that of molluscs or insects, for instance, because the ways of life and corresponding dominance sequences are grossly different.) A major category might be that of self-propelled, unattached organisms of medium to large size: vertebrates are dominant here; man is a vertebrate. Among these, terrestrial forms subdivide the major way of life: mammals dominant; man a mammal. Among land mammals as a whole, rodents are now clearly and strongly dominant, but within the Mammalia man belongs definitely to a dominance sequence quite different from that of the rodents and in this sequence he has recently and decisively risen to dominance. This criterion gives no justification for considering man the highest among all the forms of life, but it places him at least among the highest in his own general and particular sphere, and that is as far as the criterion can go.

This sort of succession of dominance within particular spheres, environmental and adaptive, suggests as another possible criterion of progress the successive invasion and development of these spheres. The succession is, again, multiple and does not give a single line of progress for all of evolution even if combined with dominance within each sphere. After early, multiple, basic, adaptive successions in the water, a major step here was spread of life from water to land, a step which tremendously increased the amount of life in the world and can be related to progress in the most general of all features of evolution, the total expansion of life. The step did not take place all at once or through a single group of organisms; the lines are multiple, but for each this step is from lower to higher by this particular criterion. By this criterion, too, man is a high type of animal, for his general environment was among the last to be filled by life. It really is legitimate to go beyond this and to point out that man's particular adaptive type was the latest to be developed up to now in the history of life, one radically new, never before exemplified, and with extreme potentialities for expansion. In the spreading sequence of adaptive types there is thus a criterion that seems quite definitely to place man at the top, as the highest type of organism in this particular respect.

Replacement within adaptive zones is another possible objective criterion of progress, but one that seems rather equivocal in spite of glib generalizations as to "higher" types continually replacing "lower." When replacement does occur, there seems to be pragmatic evidence that the successful group is somehow better. Presumably the "better" often or usually means improved adaptation of some sort, a special case of another and different possible concept of progress as moving toward better adaptation. The criterion of replacement is not generally applicable, and its bearing is more than dubious when, as has happened repeatedly, replacement is delayed or is successive without actual competition. The criterion does not apply to man, who, as just pointed out, is not a replacing form but one developing a totally new adaptive type.

The concept of progress as involving improvement in adaptation is rather obvious and apparently promising, for certainly adaptation and changes in it are among the most nearly universal features of evolution; but on closer examination it seems to be quite limited as a criterion for progress and of little significance in an attempt to establish any very general idea of progress in evolution. It is another criterion that gives multiple progressive sequences, and a really excessive number of these. Every organism is adapted to some particular way of life. This has always been true and always will be. Evolution does not proceed from the general to the particular but from the particular to the particular, building up and filling up the general as it goes. On the basis of adaptation alone there is no reason to consider one adaptive type higher or lower than another. Diversity of adaptive types has increased. This is the expansion and succession already discussed—but there is no way to decide that one adaptive type is more adaptive than another.

Within each of the myriads of distinct adaptive types there is a possibility for improvement in efficiency and for sharpening of the adaptation, narrowing its scope. In evolution this is the widespread phenomenon of increasing specialization. This, too, is progress of a sort and if we care to define it so; at least, it is a very frequent occurrence in evolution, one that is progressive in the sense of succession and trend, and one that does tend, on the whole, to increase the total of life, as summed over all its various lines. If we feel intuitively that specialization is not *real* progress, it is mainly because this is not the line our ancestry took and because it so often leads to extinction. Extinction is nonprogress, to be sure, but it may be preceded by progress, and specialization is an essential feature of the expansion of life which can hardly be denied the name of progress if progress be considered a possibility in evolution. Specialization is a form of evolutionary progress, one of many and one distinctive and peculiar in its aspects.

The peculiarity of progress in specialization resides mainly in its inverse relationship to the possibility of further progress. Change toward increased specialization approaches a limit and tends in general to decrease the possibility of change in any other direction, although the relationship is not rigid or invariable. Now the possibility of further progress has been but (in my opinion) should not be used to define progress going on or accomplished in the past. If potentiality for further advance were a sign of progress, then of course the first living thing would be the most progressive organism that ever lived! This is absurd, and so is the commonly stated conclusion that specialization is not "real" progress because it limits further progress. Yet progress that does not narrow, or that broadens, the chances of further progress is evidently a different thing from progress that is self-limiting, as specialization often is.[4] The difference becomes important only in connection with an actual movement of evolution that shows progress on other criteria than those of limitation or expansion of possibilities for future change.

There is a progress in adaptability which often tends to conflict with or be restricted by progress in adaptation (specialization), but which only comes into play if, in fact, further adaptation does occur and that adaptability is utilized in a way that can be considered progressive. That intricate sentence really does mean something, and the meaning is particularly significant for the human species. Our ancestors did not progress much as regards specialization. By that criterion, we stand low among mammals and indeed rather low among animals in general. This is the only criterion of progress that we have found by which man is a low type of animal. In fact in our rise there was a broadening and not a narrowing of the specificity of our adaptation. Our own activities, with the invention of clothing, tools, and the rest, have still further broadened that specificity and have placed us beyond any other organism in adaptability. The exploitation of that adaptability has occurred, and is still going on, as man adapts further not to one but to a great variety of ways of life. This is progress of a sort and to a degree altogether unique in the history of life.

This development is often expressed as, or I might say "confused with," progressive independence from the environ-

4. This has been recognized in various terms and to various degrees by all evolutionary biologists from Darwin onward. It has been especially emphasized and elaborated by J. S. Huxley in many studies, several elsewhere cited in the present book. It is amusing to find some anthropologists later claiming this as their discovery and criticizing biologists for *failing* to recognize it! For example: M. D. Sahlins and E. R. Service, *Evolution and Culture* (Ann Arbor, Univ. of Michigan Press, 1960).

ment. It is rather the ability to cope with a greater variety of environments than lessening of dependence on environment as a whole, and the distinction seems real and important. It is also something peculiarly human in its degree and even, in considerable part, in its kind and so must be suspect as a human and not a general or objective criterion of progress in evolution as a whole. The nonhuman example most often given for progress as increasing independence from the environment is the freeing of the amphibians from dependence on water as a living medium and then of the reptiles from dependence on water in which to lay their eggs. It has been remarkably overlooked that this was only an exchange and not a loss of dependency. Most reptiles finally became completely dependent on dry land—and water is much more widespread!

Control over the environment is still more clearly progress if we agree to consider progress frankly as defined from the human position. At least it leads us to the position that only man is really progressive in the history of life. Actual *control* of environment, as opposed to the ability merely to move about in search of suitable environments, means of escape from unsuitable ones, or the ability to get along in varied and varying environments, is almost exclusively a human ability. Such things as beaver dams are negligible in comparison and they are isolated marvels in the animal kingdom that will hardly be considered as criteria for evolutionary progress without reference to man. Means of *protection* from the environment are practically universal among organisms, although they vary greatly in nature. They may exemplify progress, on some other standard, but they do not help particularly to define it. The fact that control of environment is so nearly exclusive to man does not mean that it is not progress, but only that it is a peculiarly *human* sort of progress, part of the larger wonder that man is a new sort of animal that has discovered new possibilities in ways of life—and this is progress whether referred specifically to the human viewpoint or not.

However progress may be defined among such broader ways as have now been exemplified, there are accompanying structural and physiological changes that constitute the concrete concomitants of such more abstract progress or that are the material means by which progress occurs.

Increasing structural complication at once comes to mind, especially as evolution has so often and so misleadingly been generalized as just a succession from simple to more complex forms of life. This was an aspect of progress, and an important one, far back in the days of fundamental progress in successive occupation of the major and more radically different ways of life. The first step in this beyond the protozoans was the rise of many-celled animals, which was progress involving complication of structure. Complication was also markedly involved in the rise of the diverse multicellular phyla, but again we find that this was progress in multiple lines, not a single process of increasing complication nor even a central line with branching blind alleys that might or might not be progressive in this respect. In a broad way complication was involved in successive spread to new spheres of life, and higher forms, by any acceptable criterion, are on the whole more complex than lower; but within this broad picture correlation of complication and progress becomes quite irrelevant as regards particular cases. It would be a brave anatomist who would attempt to prove that Recent man is more complicated than a Devonian ostracoderm. Perhaps he is, but I do not know just how you would set about proving it (other than by rigging your definition of "complication"), or what it would mean if you did prove it. Yet everyone will agree, by almost any acceptable definition of "progress," that man has progressed a great deal further than the ostracoderm had.

As a matter of fact within such groups as show, for instance, the phenomenon of successive dominance, it is at least as usual for progress (by the criterion of dominance, for example) to be accompanied by structural simplification as by complication. This observation is the basis of "Williston's law."[5] As Williston, one of the great American vertebrate paleontologists, phrased it, "The parts in an organism tend toward reduction in number, with the fewer parts greatly specialized in function." This sort of change is, on the whole, in the direction of simplification rather than complication. Of course the specialization of remaining parts, as these become fewer, may complicate their individual structure, but there is no constancy in this. The teeth of cats, for instance, have become fewer and the remaining teeth have become greatly specialized in function, but these specialized teeth are considerably simpler in form than were the corresponding teeth of Paleocene carnivores.

The "law" applies in many sequences that show progress under any usual definition and this shows that progressive simplification, as well as progressive complication, may accompany progress. (It would not be valid in either case to say that such a change *constitutes* progress.) Such apparent contradiction as exists here disappears when the "law" is related to function, a fact that seems not to have been very clearly brought out in morphological evolutionary studies seeking to generalize the phenomena of structural change without reference to function. "Williston's law" is followed *only* when multiple parts combine to perform the same or closely related functions. The fewer parts continue to perform these same functions, or more specialized functions evolved from these. This particular sort of structural re-

5. Such a "law" is, of course, really just a description of what is observed to happen as a rule, or at least with some frequency, in evolution, and like most eponymous laws it was known in one form or another long before its eponym—Williston in this instance—pointed it out.

duction therefore does not represent reduction in function, which is usually retrogressive and not in the direction of progress. A familiar example of this process is found in the multiple legs of the earliest forerunners of crustaceans (trilobites), all performing the function of locomotion, and the many fewer legs of a modern crab, some of which perform the same function of locomotion and some the specialized func-

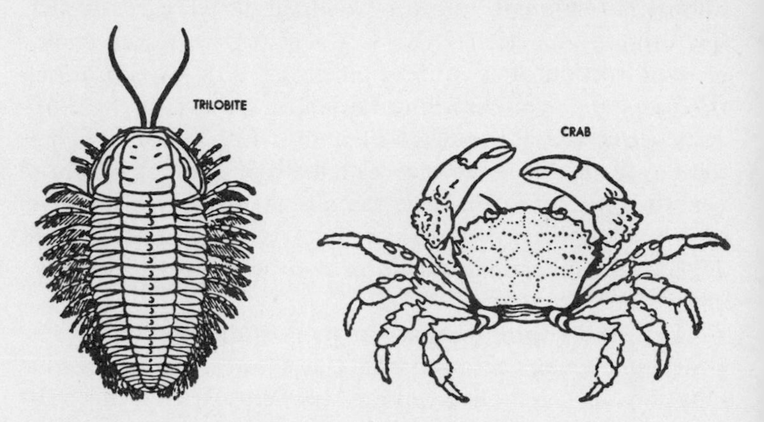

Fig. 38. An example of "Williston's law" (see text). The trilobite, an ancient forerunner of the crustaceans, has many similar segments and legs. The crab, a living crustacean, has relatively few and dissimilar segments and legs. (Data for the trilobite from Raymond.)

tion of grasping. (See fig. 38.) The very common reduction in numbers of teeth and of skeletal parts in vertebrates also occurs, as a rule, in accordance with this "law" and its limitations.[6]

6. The subject has been well discussed and examples given by W. K. Gregory in " 'Williston's Law' Relating to the Evolution of Skull Bones in the Vertebrates" (*Amer. Jour. Physical Anthropology, 20* [1935], 123–152), and in *Evolution Emerging* (New York, Macmillan, 1951).

Another group of features involved in progressive evolution and often but incorrectly used as criteria of progress arises from the ancestral-descendant relationship and relative degrees of difference from a common ancestor. It has been noted that the claimed dominance sequence invertebrate—fish—amphibian—reptile—mammal—man is *not* arranged in the objective order of dominance but in the ancestral-descendant sequence oriented with respect to man. To suppose that older groups are ipso facto lower on the scale of progress than younger groups, or ancestors than their descendants, or among contemporaries those more like the common ancestor than those less like, is to confuse change with progress. These differences denote change, no more and no less. In themselves they give no clue or criterion as to whether that change has been in the direction of progress. Of course if progress *has* occurred in the given case—something that has to be decided on other grounds altogether—*then* the older group will be lower than the younger and the ancestor than the descendant and then the amount of change from a common ancestor will reflect the amount of progress made by its various descendants.

The limitations and applications of such changes as are involved in progressive complication of structure or in "Williston's law" suggest that the concomitants of evolutionary progress should be sought in function and in structure as related plainly to function and not in the morphogenetic principles and "laws," so dear to the heart of early paleontologists, which are expressed in terms of primarily structural change. A change that seems often to be involved in progress, although evidently not perfectly correlated with it, is increase in the general energy or maintained level of vital processes. This is rather vague, but it is easy to find concrete examples. A good one occurs in the transition from reptiles to mammals, which was clearly in the direction of evolutionary progress by most definitions. The metabolic system of reptiles has a low vital minimum above which it can rise for bursts of quick but not of sustained energy, and its action is rather closely circumscribed by external conditions. The mammalian system (typically) has a higher vital minimum which is maintained at almost constant level and from which can be developed not only quick but also extraordinarily sustained energy expenditures. Many structural changes are correlated with this. For instance, the mammalian system requires effective utilization of a more constant food supply, and this in turn has been achieved by changes in the digestive system and in the means of obtaining food, including the teeth and jaws.

With regard to energy level, mammals as a whole stand near but not quite at the top among animals; among vertebrates, the birds exceed them, but mammals and birds do not belong in the same logical progress sequence by most criteria. There has been some, but not much or constant, progression in this respect within the Mammalia. Man is simply a typical mammal as far as general vital energy and its maintenance are concerned, which is to say that he stands very high among animals in general.

This particular change has carried with it other concomitants some of which are related to other rather general functional progressions. Perhaps the most important of these has to do with reproduction and care of the young. There is a sequence from mere emission of eggs and sperm into the environment, so that even mating is external and at random, through internal fertilization, to various degrees of care for the fertilized eggs until the young are born, and finally to postnatal care for the young. The sequence, or parts of it in a great variety of forms, occur in numerous different evolutionary lines which are progressive, by most different definitions of that term. It reaches a very high development among birds, which are able by their constant and high body temperature to provide a nearly uniform and favorable environment for the developing egg and which provide considerable postnatal care to the young. The mammals are, however, the highest animals in this particular respect and the case is clear-cut and indisputable, from the protection and uniformity of internal gestation through most highly perfected postnatal care including provision of nearly uniform and highly nutritious food from the mother. Man has added to this the longest and most elaborate postnatal care among the mammals and so, in this sequence, is decisively the highest of all animals.

Reproductive method is particularly closely related to natural selection, which is differential reproduction, and progression in this respect has far-reaching evolutionary consequences beyond the more immediate fact that it makes production and survival of the young increasingly certain. Internal fertilization makes possible choice of mate and introduces a new and important adjunct of natural selection. Care of the developing egg and provision of increasingly constant environmental conditions for it minimize environmentally induced variation, which is not hereditary and not directly subject to control by natural selection, and present for such control variations more strictly hereditary. Postnatal care of the young leads (even in nonhumans) to education and to a new sort of heredity, the transmission of learning, which again reaches a unique level in man.

Another and probably even more important element in many lines and sorts of evolutionary progress has been "change in the direction of increase in the range and variety of adjustments of the organism to its environment," as Herrick has put it.[7] Note well that this is not at all the same thing as "increased control over and independence of the environment." It is increased awareness and perception of

7. In a very short note which is worth many pages of less pithy discussion, including, I fear, the present pages: C. J. Herrick, "Progressive Evolution" (*Science, 104* [1946], 469).

the environment and increased ability to react accordingly. (I know that many students of the subject deny that any animals below man, or certainly the really lowly types, are "aware" or "perceive" anything whatsoever; for the present purpose, this does not matter in the least; all animals *act* as if they had awareness and perception—even the ameba does.) The progressive trend is to gather more and different kinds of information about the environment in which the organisms do, in fact, exist and to develop apparatus for appropriate adjustments in accordance with this information. On lowest levels this involves only diffuse sensibility and reactability to such signals from the environment as do have an effect on any protoplasm: some types of radiation, motion and bodily contact, temperature, and chemical effects. At higher levels it involves very complex and specialized sensory organs and equally or more complex nervous systems and other co-ordinating mechanisms.

Earlier discussion of photoreceptors has already exemplified the kind of structural progression that may be associated with this sort of functional advance. It showed, too, how progress may be multiple and not in one line or toward a single type of perfected sense organ, even when the environmental signals to be received are all of the same sort. Environments differ in the stimuli that are present, and the usefulness of sensations of various sorts also differs in different ways of life even within what is, otherwise, the same environment. Thus the lateral line organs of fishes, which are receptors for variations in pressure or movement in a surrounding liquid medium, are lost in terrestrial animals which do not live in such a medium, but the related ear mechanism is adapted for reception of rhythmic pressure variations (sounds) in air. This mechanism shows marked progress, and in general there is progress, eventually very great, in the central control and reacting mechanisms. Pit vipers, although they live in environments similar to those of other snakes and broadly of many other terrestrial vertebrates, have in the pits for which they are named a special sense organ which is a directional receptor for low degrees of heat radiation, a mechanism found in no other animals. The apparatus detects the body heat of small mammals on which these vipers live. This peculiar development certainly counts as progress in perception (and the appropriate reactions are also there, of course), but the absence of such an apparatus has no relevance to progress in animals that do not live on warm-blooded prey or that have other adequate means of locating such prey.

This sort of progress, with its innumerable different degrees and many different lines of progression, is more widespread and also more fundamental than such features as independence or control of environment. It is, indeed, progress in perception of and reaction to environment that underlies and makes possible such quite limited degrees of independence and control as have been achieved. It is also evident that this concept of progress is not man-centered but provides a general criterion applicable without necessary reference to the human condition. As a matter of fact, when the criterion is applied to man, it is another that places him at the highest level of evolutionary progress. No organism has receptors and analyzers for all possible signals from the environment, or for more than a small fraction of these. For the stimuli actually present in his environment man has about as good a set of receptors as any animal. His associated perceptual, coordinating, and reacting apparatus is incomparably the best ever evolved. Moreover in the new sort of evolution so characteristic of man, that of social structure and transmitted learning, man has supplemented his organic receptors with external, inorganic receptors of a range and delicacy unknown to any other animal—such as the radio, to mention only the most familiar of many examples.

Progress in awareness and reactability to more numerous degrees and more different kinds of environmental stimuli is linked with adaptability and often carries with it progress in individual versatility. This increase in the possible range and variety of reactions by the individual organism makes each more independent as a unit and more distinctive in its particular reactions and interrelationships. There may be, in short, an accompanying progress in individualization. This, too, is widespread in the various evolutionary lines among animals, although like other sorts of progress it is far from universal. (The virtual absence of progressive awareness in plants can be directly related to their relative lack of individualization.) Everyone has noticed that a mammal or a bird is much more individualized than, say, an oyster, not in structure or appearance, which may be more varied and more distinctive of the individual among oysters than among mammals or birds, but in reaction patterns, behavior, or general ability. Progress in versatilization of the individual and in individualization within the species has been carried by human evolution to altogether new heights. Individualization is a prerequisite for the human type of socialization and finds in the latter opportunity for its greatest possible development—an apparent paradox that is, in fact, readily resolved and has the greatest importance in understanding of man's evolutionary status, a subject to be discussed in the last sections of this book.

In summary, evolution is not invariably accompanied by progress, nor does it really seem to be characterized by progress as an essential feature. Progress has occurred within it but is not of its essence. Aside from the broad tendency for the expansion of life, which is also inconstant, there is no sense in which it can be said that evolution *is* progress. Within the framework of the evolutionary history of life there have been not one but many different sorts of progress.

Each sort appears not with one single line or even with one central but branching line throughout the course of evolution, but separately in many different lines. These phenomena seem fully consistent with, and indeed readily explained by, the naturalistic theory of evolution sketched in the last chapter. They are certainly inconsistent with the existence of a supernal perfecting principle, with the concept of a goal in evolution, or with control of evolution by autonomous factors, a vital principle common to all forms of life.

As J. S. Huxley has defined progress[8] it involved increasing complexity, control over the environment, independence from the environment, individualization (or "individuation"), and capacity for acquiring and organizing knowledge, for experiencing emotion, for exerting purpose, and for appreciating values. As Huxley points out, these are not universal trends in evolution. From first to last these progressive changes become less and less general until the final capacities specified are virtually confined to man's closer relatives or at last to man himself. In fact, this is strictly an *ad hoc* definition, that is, not a standing recognition of evolutionary progress but a specification of progressive changes that led ultimately to just one species among millions, *Homo sapiens*, and that in combination are entirely relevant only for the evolutionary status of that single species. This is not progress in a general or objective sense, and it does not warrant considering man's ancestry as the central line of evolution, which has had no central line but uncountable multitudes of divergent lines each with its own progression and possible *ad hoc* designation of progress.

8. This particular list is summarized from *New Bottles for New Wine* (New York, Harper and Bros., 1957). Somewhat different, but not contradictory, definitions are given in many others of Huxley's works, such as *Evolution in Action* (New York, Harper and Bros., 1953).

Progress leading to man was inherent in the evolutionary process only in the banal sense that it did occur, and progress leading to, say, an orchid or a flea was no less inherent in the same sense. Yet on balance of all definitions man is certainly among the highest products of evolutionary progress, and the *ad hoc* definition specific to man is after all the relevant definition for consideration of the meaning of evolution to man.

XVI. HISTORICAL RETROSPECT: THE EVOLUTION OF EVOLUTION

Man's attempts to understand the history of life have had an evolution of their own, an intellectual evolution in which progress toward truth and comprehension has been, if not constant, at least continual. The history of this progress casts light on the meaning of evolution to man. Consideration of successive contributions and of rejected alternatives shows not only how the current understanding of evolution and of man's place in it was achieved but also why and with what degree of confidence the present views on these subjects are to be received. Such consideration, necessarily summary, is the remaining necessity before turning to the final phase of this enquiry, examination of the philosophical and ethical implications of all that has gone before.

The Greeks, especially Anaximander, Empedocles, and Aristotle in, respectively, the sixth, fifth, and fourth centuries B.C., are usually credited with the first major strides toward comprehension of the history of life and specifically toward discovery of the truth of evolution. They will not long detain us here because, as a matter of fact, their contribution in this particular respect was almost negligible. They, or at least Aristotle, made some progress in the description of nature but little in the understanding of it, and indeed on the whole they placed impediments in the way of this. Aside from some speculations quite on a par with the primitive myth and magic of savages—who widely hold beliefs that contain the germs of evolutionary ideas

—the most nearly historical biological concept reached by the Greek philosophers was that of a progression of organic forms, as abstract types, from lowest and simplest to highest and most complex (man, of course). The concept was idealistic and essentially nonhistorical, for it was not related to the realistic and historical facts of organic descent through successive generations.

This metaphysical attempt at explanation of the nature and diversity of the forms of life crystallized as dogma and was the most serious obstacle in the way of inductive and truly scientific study of life for many centuries. The idealistic progression was interpreted theologically as the pattern of creation and its realistic manifestation as the static result of special creation by God of each kind of living thing as it now exists. In the light of the extremely limited knowledge of life processes prior to about the seventeenth century and practically complete ignorance of the rock record of the history of the earth, this was a reasonable hypothesis and could be considered as such from the scientific as well as from the theological point of view. In retrospect it seems entirely obvious that there was an equally reasonable alternative hypothesis, that the progression was an idealization of a material historical sequence. It even seems that there was really no legitimate theological issue at stake, for creation according to static pattern is neither more nor less theistic than creation by historical process and the two could, and can, about equally well be accommodated to revelation and to what is really essential in Christian or other religious dogma.

The hypothesis of special creation did, nevertheless, become so deeply involved in Christian belief that it was felt to be essential to that belief, and still is by a considerable proportion of Christians. Thus in the viewpoint of the church and in the emotions of its members it could not be considered as a hypothesis at all, subject to investigation and verification and to consideration of alternative hypotheses on the same basis. To accept such a view would have threatened the emotional and not infrequently also the physical and economic security of the true believers. In such an atmosphere conscious formulation of the otherwise so obvious alternative hypothesis was subjected to psychological repression in the individual, and if it had risen to consciousness its open expression would have been subjected to still fiercer social repression. In fact no such expression was made during the first fifteen centuries of the Christian era.

It is hard to believe that no solitary thinker ever thought of the historical hypothesis during this long time when the truth, far from prevailing, had no chance for expression or serious consideration. That such lonely gropings may have occurred is suggested by some evidence that sub rosa consideration of the historical hypothesis may have been rather widespread in the seventeenth century, when the psychological and social sanctions against it were beginning to relax, although still strong enough to prevent fully overt consideration of this topic. There are, for instance, definite but timid hints in the works of Descartes (who died in 1650) that he privately held evolutionary views. More open statement was suppressed by "distaste for hell-fire and respect for the church."[1]

The subject was about to break into the open, not as an opposing dogma or an anticlerical movement, but as a possibility regarding the facts of existence on the earth subject to human investigation and eventual decision. The honorable list of the pioneers who finally gave it this status would be too long here, but some names must not be omitted: Hooke (1635–1703), Ray (1627–1705), de Maillet

1. A summarized explanation of Descartes' determination not to publish any of his scientific discoveries, as given in the analytical table of contents of E. T. Bell, *Men of Mathematics* (New York, Simon and Schuster, 1937).

(1656–1738), Maupertuis (1698–1759), Buffon (1707–88), and Erasmus Darwin (1731–1802, grandfather of Charles). All of these openly considered evolution as a possibility and several of them proclaimed it as a reality, at least in a limited sense. They were, however, unable to produce from the still scanty store of facts then available sufficient weight of evidence to overcome the inertia of centuries and the almost unanimous opposition of theologians, whose grounds were not scientific and are therefore not hereafter pertinent to the history of evolutionary science. They also failed to produce a general and consistent theory as to *how* evolution works (which, of course, is quite a different matter from *whether* it works), although they had ideas on this subject which in one or two instances—as later was found—came close to the mark.

Lamarck (1744–1829) holds an important place in the history of evolutionary theory because he explicitly concluded that evolution is a general fact embracing every form of life in a single historical process, and he proposed an inclusive theory as to how this process works. It happens that his theory did not stand up in the light of later knowledge and no one follows it today, not even those who call themselves "Neo-Lamarckians," if any such remain. It was, nevertheless, a brave attempt. Lamarck took over, as was practically inevitable, the old idea that all the forms of life represent a progression with man as its highest form. He was, however, too well acquainted with existing plants and animals to fail to notice that this is not really true. He therefore concluded that although this progression is the major feature of evolution and its principal cause (but the progression's cause not specified), it is perturbed and does not work out regularly because of local adaptation to so many different ways of life. Such adaptation he supposed to be brought about by the habits of animals. The environment would require habits appropriate to it, the habits would produce structural changes from use or disuse of the various organs, and these changes would be inherited.

We know now that changes arising from habits cannot be inherited, except in the quite different and characteristically human inheritance of learning, and that there is no such thing as the old Aristotelian single progression from simple and imperfect to complex and perfect. Yet Lamarck observed, more clearly than many of his successors, that evolution involves both oriented and apparently nonoriented features and that both must be explained in a consistent general theory. He also focused attention on adaptation, an evident and universal fact in nature, as a problem that *must* be faced in any attempt to explain evolution. This was all the more true in Lamarck's day and through Darwin's generation because the then numerous supporters of special creation as a scientific theory, not as dogma or divine revelation, all pointed to adaptation as the one unanswerable argument in favor of their view.

It is not, then, surprising that the main theme of the work of Charles Darwin (1809[2]–82) was not establishment of the truth of evolution, although he did that too, but an attempt to explain adaptation and to erect a general theory of evolution on the basis of this explanation of adaptation. Darwin's whole theory, in its final development at his own hands, involved four major factors which were, in order of his opinion as to their decreasing importance: natural selection, inherited effects of use and disuse (as suggested by Lamarck), likewise inherited direct action on the organism by external conditions (a factor denied by Lamarck but

2. Here is an interesting coincidence that I have long cherished, not because it means anything but because it permits remembrance of three important dates for the price of one: Lamarck's most important theoretical work, the *Philosophie zoologique,* was published in the year of Darwin's birth, and Charles Darwin and Abraham Lincoln were born on the same day.

emphasized by his also pre-Darwinian successor St. Hilaire), and finally, "variations which seem to us in our ignorance to arise spontaneously." Of these, natural selection is particularly associated with Darwin's name to such an extent that it is commonly supposed to be the whole of the Darwinian theory, of which it was the most important but not the only part. Adaptation was explained in the main by natural selection, as it still is.

Darwin was one of history's towering geniuses and ranks with the greatest heroes of man's intellectual progress. He deserves this place first of all because he finally and definitely established evolution as a fact, no longer a speculation or an alternative hypothesis for scientific investigation. His second greatest achievement was correct identification of a major element in the rise of adaptation: natural selection. It must, however, be noted that the modern concept of natural selection has been considerably broadened and refined and is not quite the same as Darwin's. He recognized the fact that natural selection involves differential reproduction, but he did not equate the two. In the modern theory natural selection *is* differential reproduction, plus the complex interplay in such reproduction of heredity, genetic variation, and all the other factors that affect selection and determine its results. In the Darwinian system, natural selection was mainly elimination, death of the unfit and survival of the fit in a struggle for existence, a process included in natural selection as it is now known but not forming all or even the major part of that process.

Darwin's views as to the operation of evolution had other deficiencies which were flaws not in his genius but in the knowledge of his time. Chief among these was almost complete ignorance of how heredity actually operates and how hereditary variations arise. This and the limited view of natural selection as involving only the death or survival of individuals whose variations were taken as given, raised some doubt as to whether his theory really accounted for the *origin* of species and not merely for the elimination of some variations. His explanation of adaptation also came under fire with two oddly contrasting arguments: it was said on one hand that adaptation is too universal and too refined in detail to be accounted for by mere weeding out among random variations, and on the other hand that adaptation is not universal enough and that nonadaptive, still more inadaptive, characteristics should have been weeded out if natural selection were really an evolutionary factor. In other words, it was insisted that evolution should be all random or all oriented, and the apparent mixture of the two is inexplicable or at least unexplained.

While Darwin's demonstration that evolution *has* occurred was encountering the inevitable attacks based on prejudice, emotion, and dogma, his theory as to *how* it occurred was thus also attacked by fellow evolutionists and on properly scientific grounds. An opposing school developed, based essentially on the universality of adaptation and the extremely intimate and pervasive relationship between structure, function, and environment. It was felt, although the necessity was never really logical, that this *must* imply a direct and causal relationship among the three. Exponents of this view (prominent among whom were the French Giard, 1846–1908, and the American Cope, 1840–97) took the second and third elements in Darwin's theory, eliminating the rest, especially natural selection, and developed them into an antithetical theory which, peculiarly, became known as Neo-Lamarckian. The designation is peculiar because their theory altogether omitted the central point of Lamarck's views, that of a general perfecting tendency in evolution, and particularly stressed direct action of the environment on organic structure, which was explicitly denied by Lamarck. They did, however, also include effects of use and disuse, which had been emphasized by Lamarck and, among others, by Erasmus Darwin before him and to less degree by Charles Darwin after him. This theory, like the one that was truly Lamarckian, is now a historical curiosity. Unsatisfactory in many minor particulars, the theory had one major and vital error: it assumed that acquired characters, those induced by environment and habit, are inherited in kind. This assumption was universal in the nineteenth century; Darwin also accepted it without serious question. It was, however, false, as was quickly and conclusively proved with the rise of the young, still new science of genetics. Without this basis, the Neo-Lamarckian theory cannot possibly work.

Although now defunct, this theory had a strong influence on further progress in the study of evolution and contributed significantly to present understanding of the subject. Its emphasis on the relationship of structure-function-environment brought out a wealth of pertinent facts and made plain that the forces of evolution must effectively integrate the three. It also made evident some defects in Darwin's theory that further progress in knowledge had to eliminate and it particularly exposed the untenability of the extreme position taken by some of Darwin's followers, the Neo-Darwinians.

The Neo-Darwinians, centering especially around Weismann (1834–1914), out-Darwined Darwin. They rejected all of Darwin's own theory except its principal element, natural selection (in the strictly Darwinian, not the modern sense), and claimed this as essentially the whole process of evolution. In so doing they piled up much further evidence that natural selection does, in fact, operate and acquired considerable new insight into how it operates. They thus helped greatly to lay the firm basis on which rests our present knowledge of evolution. But by taking such a narrow, almost ridiculously oversimple view of evolution, they were forced into an untenable position.

During the latter part of the nineteenth century and into

the twentieth these two opposing materialist schools of evolutionary theory lambasted each other, the Neo-Lamarckians demonstrating that Darwinian natural selection cannot be the whole story, the Neo-Darwinians demonstrating that it must be part, at least, of the story and that acquired characters are not inherited as such. Each side, of course, had part of the truth and between them they had the whole truth almost within their grasp had they only been able to see this. Much of the trouble was that they fought on the wrong issues and asked each other the wrong questions. The correct issue on natural selection was not whether this is or is not the cause of adaptation, but how adaptation arises in the interplay of multiple forces of which Darwinian natural selection is obviously one, but only one of many. Similarly the issue on inheritance of acquired characters was put in the wrong terms. Characters, as such, are not inherited, whether they be acquired characters or not. It is a series of determiners for a developmental system that is inherited. What characters result from this depends on the interplay of the inherited determiners, the activities of the organism, and the environment during development. Thus the Neo-Lamarckians were right that there is a functional and causal relationship here, but the Neo-Darwinians were right that its results are not inherited *as such*.

Failure to grasp the essence of these issues and failure to put together their pieces of the truth into a whole was largely caused by the fact that a large piece was still missing. This was later to be supplied by the geneticists.[3]

3. That Darwin's ideas about inheritance were almost completely wrong has been considered to vitiate his theories of evolution, and some geneticists have even maintained that it was impossible to have any valid theory of evolution until the recent rise of molecular genetics. On the contrary, given the facts already indisputable in Darwin's day that variation occurs in all populations and that variations are inheritable to considerable degree, it is relatively unimportant for evolution at the

In the meantime, many students of evolution who were not rabid devotees of one of these two schools or the other fell into moods of suspicion or despair. There were two main, ostensibly naturalistic theories and you could take the proof supplied by each that the other was untenable. What, then, to do? One could gather more facts and suspend judgment as to what meaning they might eventually have. This sounds like a fine idea and it has even been hailed (following Bacon, who never succeeded in making a discovery by his famed inductive method) as the proper procedure in science. In reality, gathering facts, without a formulated reason for doing so and a pretty good idea as to what the facts may mean, is a sterile occupation and has not been the method of any important scientific advance. Indeed facts are elusive and you usually have to know what you are looking for before you can find one.

If unwilling merely to suspend judgment, one could try to find an acceptable alternative to the two main naturalistic theories of evolution. The trouble with this was that both were about as much right as they were wrong and efforts to avoid them both tended merely to be all wrong. There was no real success, even partial, in this effort until Mendelian and then population genetics came along and even this, as we will see, confused the issue still more before it clarified it.

Or one could throw over naturalistic, materialistic, or causalistic theory altogether. It was in this atmosphere that

---

organismal, Darwinian level to know how the inheritance is carried at the subcellular level. See, for example, E. Mayr, "Darwin and the Evolutionary Theory in Biology," in *Evolution and Anthropology: a Centennial Appraisal* (Washington, Anthrop. Soc. Washington, 1959). Of course eventual knowledge of the *whole* evolutionary process must include everything from the molecular (or even the atomic?) level upward, but, as I previously noted, the recent triumphs of molecular biology have as yet contributed comparatively little of essential value for understanding of organismal, evolutionary biology.

most of the extremely diverse vitalistic and finalistic theories, a veritable spate of them, were advanced. In almost all of these a sense of despair or of hope, an emotion even more blinding than despair, is evident. The relatively few first-hand investigators of evolution who abandoned causalism did so, for the most part, because they despaired of finding an adequate naturalistic theory and could not endure the void of having no theory at all. Others, among them a number of professional and amateur philosophers, sounded a note of hope which was often quite plainly hope of drawing meaning from something not understood or, and this is particularly striking, hope of finding that science did, after all, confirm what were in reality their intuitive or inherited and popular prejudices. This tendency to confirm prejudice accounts for the great popularity that finalist theories have sometimes enjoyed among those incompetent to judge them adequately from either a scientific or a philosophical point of view.

All these theories, vitalist, finalist, or both, involved some degree of abandonment of causalism. They did not explain evolution, but claimed that it is inexplicable and then gave a name to its inexplicability: "élan vital" (Bergson), "cellular consciousness" (Buis, under the pseudonym "Pierre-Jean"), "aristogenesis" (Osborn), "nomogenesis" (Berg), "holism" (Smuts), "entelechy" (Driesch), "telefinalism" (du Noüy), "the Principle of Organization" (Sinnott), "Omega" (Teilhard)—the list could be greatly extended. As Huxley has remarked, the vitalists' ascribing evolution to an *élan vital* no more explained the history of life than would ascribing its motion to an *élan locomotif* explain the operation of a steam engine. The finalist left explanation still further behind, for he did not render even such lip service to causality as was often made by the nonfinalistic vitalists. In many cases the finalist merely viewed the phenomena of life with the unreasoning wonder of a child and decided that they happened simply because they were *meant* to happen. In other cases, as an eminent student remarked in a different context, "The finalist was often the man who made a liberal use of the *ignava ratio,* or lazy argument: when you failed to explain a thing by the ordinary process of causality, you could 'explain' it by reference to some purpose of nature or of its Creator."[4]

The general features of vitalist and finalist philosophy and theory have already been outlined on previous pages. Once causalism is abandoned, there are no limitations on flights of the imagination and there are about as many separate vitalist and finalist theories as there have been vitalists and finalists. A review of each of these individually would be merely tedious and is not necessary. It has been shown that although the basic propositions common to all of them are nonscientific, that is, defined as outside the limits of scientific investigation, these would necessarily involve phenomena that can be investigated. It has also been pointed out that these diagnostic phenomena are in fact absent in the history of life. The mere fact that vitalism and finalism do not *explain* evolution would not warrant concluding that they therefore are not true: one cannot logically exclude a priori the possibility that evolution might really be noncausalistic and inexplicable. But the fact that the history of life is flatly inconsistent with their basic propositions does warrant the conclusion that vitalism and finalism are untenable.[5]

4. P. G. Galloway, *Philosophy of Religion* (New York, Scribner, 1914).

5. And the previous demonstration of this fact warrants the present unfavorable review of the position of vitalism and finalism in the history of the study of evolution. Most of the literature of this subject is either ingenuous or ingenious special pleading and little of it can be recommended to a reader who may want to read a more favorable and still a judicious review. Fortunately there is one such work, although it is

It must be concluded that the tendency of the vitalist and finalist theories was rather to obfuscate than to advance the study of evolution. They did, nevertheless, make some contribution to this. They emphasized and firmly established the fact that evolution involves forces that are directional in nature and creative in aspect and they exposed weaknesses in some earlier attempts to identify these forces. To this extent they had an influence in the synthesis that achieved the present causalistic explanation of this particular aspect of evolution. They have also had another effect not in the advance but in the acceptance of knowledge, which should not be underestimated. To great numbers of people unlearned in the subject, the fact of evolution has been emotionally distasteful and has been rejected on this basis alone, although the rejection has often been rationalized in other terms. Vitalistic and, more particularly, finalistic theories have persuaded many of these wishful thinkers that evolution is, after all, consonant with their emotions and prejudices. For them the pill has been sugar coated. In this respect, even Teilhard's *The Phenomenon of Man,* currently the most popular but even within the finalist fold one of the least truly scientific of such efforts, may have rendered a real service. The danger is that the sugar coating may be mistaken for

---

not available in English: L. Cuénot, *Invention et finalité en biologie* (Paris, Flammarion, 1941).

Although Cuénot here avows himself a finalist, he still retains the charm and sobriety that mark the long sequence of his work. He remains critical of the various finalist theories, and he is not blind to the scientific, logical, and philosophical weaknesses of the position in which he nevertheless finds himself. The reader should, however, note and bear in mind that when Cuénot wrote this book he was evidently completely unaware of the now current naturalistic theory of evolution or of any of the crucial studies on population genetics and selection that gave impetus to this. The book does not mention even his fellow Parisian leader in this field, Teissier.

the remedy. Any value in it will fail of realization unless it is a step in the direction of taking truth straight.

It was during the period around the turn of the century, when the conflicts of Neo-Darwinism, Neo-Lamarckism, vitalism, and finalism had thrown the study of evolution into great confusion, that the science of genetics was born. Even before the rediscovery of Mendelism, de Vries had noted the sudden and random appearance of new varieties of plants. (His plants were primroses and we now know that his new varieties arose from chromosome mutations.) He decided that this, at last, was the real basis of evolution and he generalized it in his "mutation theory," published in 1901.

Scientists often display a human failing: whenever they get hold of some new bit of truth they are inclined to decide that it is the whole truth. Thus the Neo-Darwinians insisted that natural selection was the whole truth of evolution; the Neo-Lamarckians held that interaction of structure-function-environment was the whole truth; the vitalists saw the whole truth in the creative aspect of life processes; and the finalists found all basic truth in the directional nature of evolution. Similarly, many of the early geneticists, although they soon learned far more about the mechanism involved, accepted de Vries' thesis and concluded that mutation was the whole truth of evolution. Mutations are random, so it was decided that evolution is random. The problem of adaptation was, in their opinion, solved by abolishing it: they proclaimed that there is no adaptation, only chance preadaptation.

Other theories had often stumbled over the fact that there is quite plainly a random element in evolution, the nature of which had been unknown. Now the mutationists had identified the source of this random element, but their theory stumbled over the fact that evolution is not predominantly random. The vitalists and finalists were right in continuing to insist on this point, although they were wrong in their own overgeneralization of insisting that the directional element is universal and in maintaining that this element is inherent in life or in its goal. The mutationist discoveries were bewildering to many field naturalists and paleontologists, because they in particular were well aware that evolution *cannot* be a purely random process and that progressive adaptation certainly does occur. For a time the discoveries of the geneticists seemed only to make confusion worse confounded. Defeatism and escapism spread among many students of evolution. One eminent vertebrate paleontologist (Scott) ended a lifetime of study of evolution with the conclusion that he did not, after all, know anything about its causes; another (Broom) decided in the declining years of his prolonged and exceptionally fertile studies of the subject that good and bad angels must be directing evolution!

In fact, as the geneticists' studies progressed they were providing the last major piece of the truth so long sought regarding the causes of evolution. As this went on, it naturally began to dawn on students of the subject that each of the conflicting schools of theory had part of the truth and that none had all of it. The movement began with geneticists who wondered what effect natural selection, if it should really be a guiding force in evolution, would have on the genetic factors that they had discovered in individuals and in populations. They found that it would have a profound effect and that this effect was not exactly that predicted either by the Neo-Darwinians or by the mutationists. These initial successes intensely stimulated study of evolution, which quickly regained its slipping position as the focal point of all the life sciences. Students of all the different aspects of life began to unite, each contributing from his field its special bit of evolutionary fact to add to the growing synthesis. The resulting synthetic theory[6] need not

6. The theory has sometimes been called Neo-Darwinian, even by those who have helped to develop it, because its first glimmerings arose from confrontation of the Darwinian idea of natural selection with the

here be summarized, because it is the theory already presented, in broad outlines, in the preceding chapters.

The synthetic theory has no Darwin, being in its nature the work of many different hands. To mention any of these is to be culpable of important omissions, but if only to indicate the breadth of the synthesis it may be noted that among the many contributors have been: in England, Fisher, Haldane, Huxley, Waddington, and Ford; in the United States, Wright, Muller, Dobzhansky, Mayr, and Stebbins; in Germany, Timoféeff-Ressovsky and Rensch; in the Soviet Union, Chetverikov and Dubinin; in France, Teissier; in Italy, Buzzati-Traverso.[7] I do not, of course, mean to say that these students all hold opinions identical in detail. Their fields of work, not to mention other personal variables, are so diverse that this would be a miracle. All, however, have made outstanding contributions to the modern synthesis and all seem to be agreed as to its most essential features.

This work has placed the study of the causes of evolution on a new and firm footing and has produced a degree of agreement as to these causes never before approached. We

---

facts of genetics. The term is, however, a misnomer and doubly confusing in this application. The full-blown theory is quite different from Darwin's and has drawn its materials from a variety of sources largely non-Darwinian. Even natural selection in this theory has a sense distinctly different, although largely developed from, the Darwinian concept of natural selection. Second, the name "Neo-Darwinian" has long been applied to the school of Weismann and his followers, whose theory was different from the modern synthetic theory and should not be confused with it under one name.

7. Those named are just a few who contributed basically to the theory in its most crucial formative years, roughly 1930–1945. Since then work in this field has been so extensive and so intensive that hundreds of names of significant workers could now be added. For what it is worth, my own work directly within this body of theory began in 1935.

seem at last to have a unified theory—although a complex one inevitably, as evolution itself is a complex interaction of different processes—which is capable of facing all the classic problems of the history of life and of providing a causalistic solution of each.

This is not to say that the whole mystery has been plumbed to its core or even that it ever will be. The ultimate mystery is beyond the reach of scientific investigation, and probably of the human mind. There is neither need nor excuse for postulation of nonmaterial intervention in the origin of life, the rise of man, or any other part of the long history of the material cosmos. Yet the origin of that cosmos and the causal principles of its history remain unexplained and inaccessible to science. Here is hidden the First Cause sought by theology and philosophy. The First Cause is not known and I suspect that it never will be known to living man. We may, if we are so inclined, worship it in our own ways, but we certainly do not comprehend it.

Within the realm of what is clearly knowable, the main problem seems to me and many other investigators to be solved, but much still remains to be learned. Our knowledge of the material history of life is considerable, but it is only a tithe of what we should and can know. Study of the intricate dynamics of natural populations and their interrelationships is well begun, but not much more. Study of the evidently very important psychological factors in evolution has just begun. In the study of individual life processes are gaps that are probably the most serious remaining in general knowledge of evolution.

It is to be expected that future discoveries will not only greatly deepen but also modify our current ideas of evolutionary processes. It is, however, improbable that these ideas will be vitiated in an essential way. If we do not know each single step from reptilian ear to mammalian ear, we know a multitude of steps along the line. If we do not know how genetic system leads to organic system, we know that it does and we know, broadly, the equivalents in the two. Any report on scientific enquiry or on its human import will be an interim report, as long as our species lasts and continues to value truth, but in the current knowledge of evolution we have an excellent basis for such an interim report.

PART THREE

Evolution, Humanity, and Ethics

"Out of all our study of science we should collectively by now have gained not only knowledge but also some wisdom about the meaning of things that we can apply to human relations. Clearly this wisdom must come predominantly from a consideration of the evolutionary process, for that is the process, according to the belief of virtually all scientists, whereby man developed his present powers."

R. R. Williams, in "Natural Science and Social Problems," *American Scientist, 36* (1948), 116–126.

XVII. MAN'S PLACE IN NATURE

The first grand lesson learned from evolution was that of the unity of life. One of the great ethical achievements of early Christianity and some of the other religions was recognition, in principle, of the brotherhood of man. While it refuted certain other intuitive conclusions of various Christian and other theologians, confirmation of the truth of evolution established this doctrine as a scientific fact. It also carried the conclusion to a much higher level, one rejected by some Christian theologians (although it is an intensification rather than a denial of basic Christian philosophy) and one glimpsed, however imperfectly, in some other religious systems. Not only are all men brothers; all living things are brothers in the very real, material sense that all have arisen from one source and been developed within the divergent intricacies of one process.

Man is part of nature and he is kin to all life. Yet the metaphoric "brotherhood," while a valid expression of the fact of essential unity, is evidently too strong a term if we consider it more literally and in greater detail. Man is related to every other organism that has existed, or that does or will exist on the earth, but it is perfectly obvious that he is more nearly related to some than to others. It is obvious that man is an animal and not a plant, which is a taxonomic way of saying that he is more closely related, not only in structural and functional resemblances but also in material origin, to all animals than to any plant. Among the animals it is equally obvious that, in the same sense, man is a vertebrate and not an invertebrate. Among vertebrates he is a mammal, not one of the various sorts of "fishes," an amphibian, a reptile, or a bird. Among mammals he is a primate, a member of the Order Primates and of no other of the numerous mammalian orders.

As the degrees of relationship are narrowed, the distinctions become less and their significance becomes less obvious. More detailed evidence is needed and its interpretation is more difficult. The general position of man within the animal kingdom, within the vertebrate subphylum, and within the mammalian class is absolutely established and beyond any doubt. His exact position within the primate order and his detailed relationship to each of the other primates, living and fossil, are at the present time approximately established but are not yet known beyond all question and with complete precision. We want and need to know these final details and certainly few subjects are more fascinating. These details have, however, no particular importance for the broad problems of philosophy and of ethics. The basis is established and all the essential questions are answered by the knowledge that all life is one and that within this vast unit man belongs to the particular subdivision known as the Order Primates. Such questions as whether *Homo erectus* (*"Pithecanthro-*

pus") is a direct ancestor of all, part, or none of the present mass of humanity or whether the gorilla or some other living primate is the closest surviving branch from the human ancestry need not concern us here, for they really have no essential bearing on the nature of man or on man's place in nature.[1]

The establishment of the fact that man is a primate, with all its evolutionary implications, early gave rise to fallacies for which there is no longer any excuse (and never was much) but which still sometimes affect the thinking of evolutionists and the reaction of those who have not yet faced the truth of evolution. These fallacies arise from what Julian Huxley calls "the 'nothing-but' school."[2] It was felt or said that because man is an animal, a primate, and so on, he is *nothing but* an animal or *nothing but* an ape with a few extra tricks.[3] It is a fact that man is an animal, but it is not a fact that he is nothing but an animal. (It is not a fact that man is an ape, extra tricks or no, and so, of course, all the less a fact that he is nothing but an ape.) Such statements are not only untrue

1. It is, however, interesting to note that the preceding questions have now been answered with high probability if not absolute certainty. It is very likely that *Homo erectus,* with its several variants, represents a stage ancestral to *Homo sapiens,* and it is virtually certain that gorilla and chimpanzees together (genus *Pan* of conservative classification) are the closest living relatives of man. See, for example, S. L. Washburn, editor, *Classification and Human Evolution* (Chicago, Aldine, 1963).

2. In his introduction to *Touchstone for Ethics, 1893–1943* by T. H. and J. Huxley (New York and London, Harper and Bros., 1947). This excellent and thought-provoking book is highly pertinent throughout all of the present enquiry, and especially the following parts of it.

3. It is to be feared that some evolutionists made such statements mainly "pour épater les bourgeois" and must have known perfectly well that they distort the truth. Similarly some physicists have delighted in distortion of the statistical nature of some of their constructs by presenting them in such striking statements as that a kettle of water placed on the fire often enough will eventually freeze instead of boiling, although they must have realized that this simply is not true.

but also vicious for they deliberately lead astray enquiry as to what man really is and so distort our whole comprehension of ourselves and of our proper values.

To say that man is nothing but an animal is to deny, by implication, that he has *essential* attributes other than those of all animals. This would be false as applied to any kind of animal; it is not true that a dog, a robin, an oyster, or an ameba is nothing but an animal. As applied to man the "nothing but" fallacy is more thoroughgoing than in application to any other sort of animal, because man is an entirely new kind of animal in ways altogether fundamental for understanding of his nature. It is important to realize that man is an animal, but it is even more important to realize that the essence of his unique nature lies precisely in those characteristics that are not shared with any other animal. His place in nature and its supreme significance to man are not defined by his animality but by his humanity.

Man has certain basic diagnostic features which set him off most sharply from any other animal and which have involved other developments not only increasing this sharp distinction but also making it an absolute difference in kind and not only a relative difference of degree. In the basic diagnosis of *Homo sapiens* the most important features are probably interrelated factors of intelligence, flexibility, individualization, and socialization. All four of these are features that occur rather widely in the animal kingdom as progressive developments, and all define different, but related, sorts of evolutionary progress. In man all four are carried to a degree incomparably greater than in any other sort of animal. All have as their requisite and basis the still more fundamental evolutionary progress "in the direction of increase in the range and variety of adjustments of the organism to its environment,"[4] which involves increased and

4. Herrick, *op. cit.* See *supra,* p. 257.

improved means of perception of the environment and, particularly, of integrating, coordinating, and reacting flexibly to these perceptions.

In other respects, too, man represents an unusual or unique degree and direction of progress in evolution, as was seen in discussion of that topic. He embodies an unusually large bulk of life substance and carries on a large share of the earth's vital metabolism. He is one of the dominant current forms of life, the latest to arise and now the only one within the particular dominance sequence to which he belongs. He has successfully replaced any competing type and has occupied a sphere of life or great adaptive zone which was, historically, the most recent to be entered by animals. His particular sort of progress has not, to this point, been self-limiting and leads to no obvious future blind end. He is, on the whole, the most adaptable of animals. He is about as independent of environment as any animal, or, as it may more accurately be put, is able to get along in about as wide an environmental range as any. He is almost the only animal that really exerts any significant degree of control over the environment. His reproductive efficiency is the highest in the animal kingdom, with prenatal protection at least as high as in any other animal and postnatal care decidedly higher.

Even when viewed within the framework of the animal kingdom and judged by criteria of progress applicable to that kingdom as a whole and not peculiar to man, man is thus the highest animal. It has often been remarked (perhaps again merely "pour épater les bourgeois") that if, say, a fish were a student of evolution it would laugh at such pretensions on the part of an animal that is so clumsy in the water and that lacks such features of perfection as gills or a homocercal caudal fin. I suspect that the fish's reaction would be, instead, to marvel that there are men who question the fact that man is the highest animal. It is not beside the point to add that the "fish" that made such judgments would have to *be* a man!

Is it necessary to insist further on the validity of the anthropocentric point of view, which many scientists and philosophers affect to despise? Man *is* the highest animal. The fact that he alone is capable of making such a judgment is in itself part of the evidence that this decision is correct. And even if he were the lowest animal, the anthropocentric point of view would still be manifestly the only proper one to adopt for consideration of *his* place in the scheme of things and when seeking a guide on which to base *his* actions and his evaluations of them. (The point is again only reinforced by the absurdity of supposing that an animal lower than man, or that man if he were not in fact highest, could consider the scheme of things or seek for guidance in it.) This is not, of course, to make or to accept an apology for the centering of man's attention so blindly on his own personality in the here and now that he fails in just these ways, does not consider the whole scheme of things, and does not evaluate his actions on this basis.[5]

It is still false to conclude that man is *nothing but* the highest animal, or the most progressive product of organic evolution. He is also a fundamentally new sort of animal and one in which, although organic evolution continues on its way, a fundamentally new sort of evolution has also appeared. The basis of this new sort of evolution is a new sort of heredity, the inheritance of learning. This sort of heredity appears modestly in other mammals and even lower in the animal kingdom, but in man it has incomparably fuller

5. It is this point that has been made solidly by, among others, R. L. Schuyler, "Man's Greatest Illusion" (*Proc. Amer. Phil. Soc.*, *92* [1948], 46–50), although he was carried away by his thesis (and I think misled by some of the physical scientists) to the point of denying validity also to the broader anthropocentric view of the universe, to forgetting that *man* is doing the thinking and the acting, *not* the cosmos.

development and it combines with man's other characteristics unique in degree with a result that cannot be considered unique only in degree but must also be considered unique in kind.

Organic evolution rejects acquired characters in inheritance and adaptively orients the essentially random, nonenvironmental interplay of genetical systems. The new evolution peculiar to man operates directly by the inheritance of acquired characters, of knowledge and learned activities which arise in and are continuously a part of an organismic-environmental system, that of social organization. Organic inheritance is limited in its spread in space and in time by its rigid requirements of proximity and of continuity. Learning inheritance was similarly limited in its earlier stages, and it still is in all the lower animals, not subject to the new evolution; but in modern man it has escaped these limitations. Means have been devised for recording and transmitting knowledge external to the organism, by graphic methods, at first, and now also by recordings of several other types and by wire and wireless transmission. In the new evolution we can inherit directly from ancestors dead two thousand years or from ancestors younger, perhaps, than we are and half the world away from us or from our organic kin. Our inheritance can be passed on, instantaneously or after a lapse of untold generations, to our whole species and without any necessary course of gradual spread in the population—a potentiality seldom fully realized and yet surely inherent in the new heredity.[6]

That sort of inheritance takes place largely through the most completely unique single characteristic of man: lan-

6. This concept of a contrast between the two sorts of evolution and the two sorts of heredity was earlier skillfully developed by a number of other students, among them E. W. Sinnott, "The Biological Basis of Democracy" (*Yale Review, 35* [1945], 61–73); and C. H. Waddington, "Human Ideals and Human Progress" (*World Review* [August, 1946],

guage. Although the word "language" has sometimes been loosely and misleadingly applied to forms of nonhuman animal communication, the distinction is absolute. Animal "language," which occurs also in man, is a limited, simple, nonsymbolic reflection of the animal's reaction to an immediate and concrete situation, usually purely emotive or, in psychological terms, affective. Human language is a virtually limitless, extremely complex, symbolic system capable of communicating anything sensed or experienced, including extreme abstractions and references to immediate or indefinitely remote past and future.[7] Not even the rudiments of this kind of language occur in any other now living animal. It is an absolute essential, probably the most important of all essentials, of the human condition not only because of its role in cultural inheritance but also because culture would not have arisen and would not persist without immediate communication by means of language.

Another human characteristic, related to the extent that it is fostered by the temporal symbolism of language, is the ability to foresee or predict the outcome of our actions. That is not quite unique in kind, for the behavior of many animals indicates some sort of rudimentary foresight, but it is unique in degree. Its major importance is that it makes possible and necessary human responsibility, to be further considered hereafter.

---

pp. 29–36). There are, of course, still earlier more or less explicit statements of the same idea. It is now universally recognized by philosophers, sociologists, anthropologists, and biologists and has been discussed at length among the latter by Dobzhansky, Huxley, Mayr, and many others.

7. I have discussed this at greater length in "The Biological Nature of Man" (*Science, 152* [1966], 472–478). A few references to the enormous literature of the subject are there given, and I may here add: E. Cassirer, *An Essay on Man* (New Haven, Yale Univ. Press, 1944), and E. H. Lenneberg, editor, *New Directions in the Study of Language* (Cambridge, M.I.T. Press, 1964).

Human social organization, also unique, is in a sense the basis of the new evolution, but more strictly it is the medium in which the new heredity operates and an evolving result of the interactions of that heredity with many other factors, both organic and social in nature. Within the structure of a species, including, of course, the human species, organic evolution operates in habitually interbreeding groups, which are always defined in part by geographic proximity although other ecological factors also normally play a part, and which, among themselves, have varying degrees of interchange of heredity. Within the structure of society as a whole the new evolution operates in what may be called habitually interthinking groups. These also are defined in part by geographic proximity and by geographic concomitants in the social organization, notably nationality and language. Such definition is not, however, required by the nature of the new heredity, as it is by that of organic heredity. Geographic restrictions on interthinking are still very evident, but they could be eliminated, and in the meantime other defining factors are more prominent and more complex than in the case of interbreeding. Science, the principal mechanism for producing new hereditary materials transmitted by interthinking, has made some progress (not nearly enough) in eliminating the geographic factors. Political, religious, economic, and (in interaction with organic heredity) racial factors loom large in delimitation of interthinking groups and often override geographic delimitation although the geographic social organization of nationality has some tendency to maintain a relationship.

The new evolution is itself a result of organic evolution, but it is something essentially different in kind. Although it may be semantically correct and scientifically enlightening to call both "evolution," it is extremely important to recognize that the difference in kind makes this in large part an analogy and not a straight equivalence. We may expect to find, and do find, that many general principles of evolution apply analogously to the two, but it is invalid and indeed dangerous to assume that equivalent evolutionary principles operate throughout the two and that principles discovered regarding one may forthwith be applied to interpretation of the other. Since we now know much more about organic evolution than about societal evolution, this fallacious sort of transfer is usually attempted from the field of biology to that of sociology. Examples of this will be discussed in connection with the search for a naturalistic ethic. The confusion of analogy and equivalence between the old evolution and the new lies at the root of the difficulties of that search and of some of its downright vicious outgrowths.

Organic evolution and societal evolution must, then, be not only constantly compared in their common aspects as evolution but also constantly contrasted in their differences as sharply distinct sorts of evolution, even though one is the product of the other and is, indeed, its continuation by other means. A few of the many contrasts have already been mentioned. One more, which may prove most basic of all, should be strongly emphasized. In organic evolution new factors arise as mutations without volition and without any fixed equivalence to needs or desires. Once they have arisen, their fate or further role in evolution is determinate and follows from the interplay of factors in which the organisms assist but over which among prehumans they had no semblance of control. In the new evolution new factors arise as elements in consciousness, although always somehow influenced by and sometimes directly produced from the tangled psychological undergrowths of habit, emotion, and the subconscious. They arise in relationship to needs and desires of individuals and commonly in relationship to the individual's perception and judgments of the needs and desires of a social group. Once they have arisen, their further evolutionary role is not mechanistically determinate and is subjected to the influence not only of the actual needs and desires of the group and of volitions extremely complex in basis but also of an even more complex interplay of emotions, value judgments, and moral and ethical decisions.

Through this very basic distinction between the old evolution and the new, the new evolution becomes subject to conscious control. Man, alone among all organisms, knows that he evolves and he alone is capable of directing his own evolution. For him evolution is no longer something that happens to the organism regardless but something in which the organism may and must take an active hand. The possibility and responsibility spread from the new evolution to the old. The accumulation of knowledge, the rise of a sense of values, and the possibility of conscious choice, all typical elements in the new evolution, also carry the means of control over organic evolution, which is determinate but is determined, in part, by factors that can be varied by the human will.

Man's control over either of these processes, even over the new evolution from which the means of control arise, is neither rigid nor boundless. He can neither make sure that evolution will certainly occur without deviation in a given direction at a given rate nor make a choice among all conceivable directions and rates. In both sorts of evolution he must still, to mention only these basic limitations, work with the changes that arise and disseminate them by existing mechanisms. In organic evolution he cannot decide what sort of mutation he would like to have[8] and he must await the processes of differential reproduction for the integration and spread of the mutation chosen. In the social process, although the rise of new materials is conscious, their range

8. At present he cannot, and when or if he ever can, he still must work on the given basis of the existing organism in which the change is to be induced.

is strongly limited at any given time. New knowledge and ideas do not and cannot come in exact accordance with our needs or wishes. Their spread and translation into action are also far from instantaneous, as we all know, and there is certainly no reason to think that this will ever depend on processes much more rapid than displacement against a great weight of inertia and active opposition. But we do not conclude that man cannot control floods because he cannot make rain fall upward. The infantile fantasy of becoming whatever we wish as fast as we please is simply unrealistic in a material cosmos, but this is obviously no argument against the fact that we do have a measure of conscious control over what becomes of us.[9] The intelligent exploitation of this possibility depends, of course, not only on determination of its means but also on recognition of its limits. Some attempt to consider the future possibilities will be made in the last chapter of this section.

In preceding pages evidence was given, thoroughly conclusive evidence, as I believe, that organic evolution is a process entirely naturalistic in its operation, although no explicit conclusion was made or considered possible as to the origin of the laws and properties of matter in general under which organic evolution operates. Life is materialistic in nature, but it has properties unique to itself which reside in its organization, not in its materials or mechanics. Man

9. Yet there has been a minor vogue for the idea that because we do not control social evolution completely and with no limitation as to possibilities, we therefore do not control it at all! There is perhaps here too a touch of the false analogy between the two evolutions and a feeling that because organic evolution happens, willy-nilly, to, say, sparrows, this is true also of social evolution and man. For an able but essentially fallacious exposition of this point of view, see L. A. White, "Man's Control over Civilization: an Anthropocentric Illusion" (*Sci. Monthly, 66* [1948], 235–247). This illogically defeatist attitude has not become popular, but it may still be upheld by one or two of White's students.

arose as a result of the operation of organic evolution and his being and activities are also naturalistic, but the human species has properties unique to itself among all forms of life, superadded to the properties unique to life among all forms of matter and of action. Man's intellectual, social, and spiritual natures are altogether exceptional among animals in degree, but they arose by organic evolution. They usher in a new phase of evolution, and not a new phase merely but also a new kind, which is thus also a product of organic evolution and can be no less naturalistic in its essence even though its organization and activities are essentially different from those in the process that brought it into being.

It has also been shown that purpose and plan are not characteristic of organic evolution and are not a key to any of its operations. But purpose and plan are characteristic in the new evolution, because man has purposes and he makes plans. Here purpose and plan do definitely enter into evolution, as a result and not as a cause of the processes seen in the long history of life. The purposes and plans are ours, not those of the universe, which displays convincing evidence of their absence.

Man was certainly not the goal of evolution, which evidently had no goal. He was not planned, in an operation wholly planless. He is not the ultimate in a single constant trend toward higher things, in a history of life with innumerable trends, none of them constant, and some toward the lower rather than the higher. Is his place in nature, then, that of a mere accident, without significance? The affirmative answer that some have felt constrained to give is another example of the "nothing but" fallacy. The situation is as badly misrepresented and the lesson as poorly learned when man is considered nothing but an accident as when he is considered as the destined crown of creation. His rise was neither insignificant nor inevitable. Man *did* originate after a tremendously long sequence of events in which both chance and orientation played a part. Not all the chance favored his appearance, none *might* have, but enough did. Not all the orientation was in his direction, it did not lead unerringly human-ward, but some of it came this way. The result *is* the most highly endowed organization of matter that has yet appeared on the earth—and we certainly have no good reason to believe there is any higher in the universe. To think that this result is insignificant would be unworthy of that high endowment, which includes among its riches a sense of values.

XVIII. THE SEARCH FOR AN ETHIC

Man is a moral animal. With the exception of a few peculiar beings who are felt to be as surely crippled as if the deformity were physical, all men make judgments of good or bad in ethics and morals. All feel some degree of compulsion to value and promote the good, to condemn and eliminate the bad. It requires no demonstration that a demand for ethical standards is deeply ingrained in human psychology. Like so many human characteristics, indeed most of them, this trait is both innate and learned. Its basic mechanism is evidently part of our biological inheritance. The degree of development of this, in the individual and in society, as well as the particular form that it will take are conditioned by learning processes in the family and in the wider aspects of the social structure. Man almost inevitably acquires an ethic and this responds to a deep need in any normal member of the species. The reasons for this need and the immediate elements in its fulfillment are among the most important of subjects for the study of man, but they are, on the whole, outside our present enquiry. It suffices here to recognize the fact that the need does exist and that it is itself a natural product of evolution.[1]

Through the ages ethical standards responding to this

1. In the volume by the two Huxleys previously cited, Julian Huxley has devoted special attention to the origin and development of the individual moral sense, basing himself mainly on Freudian theory. The evolutionary origin and significance of the need for ethics, or for "ethicizing," are discussed at still greater length by C. H. Waddington in *The Ethical Animal* (London, Allen and Unwin, 1960), and in earlier works there cited.

need have been supplied mainly from three closely interrelated sources: introspection, authority, and convention. The basis of acceptance has been intuitive, the feeling of rightness, without objective enquiry into the reasons for this feeling and without possible test as to the truth or falseness of the premises involved. Introspection may produce results so intensely felt and within such an emotional framework that they are considered as having sanction and force from some external, nonmaterial source, that is, they are taken to be inspired and divine in origin. They then become revelations. Such introspective revelations of a gifted few may in time be accepted by many others as valid, just as may also the introspections of philosophers who do not consider themselves, or do not claim, to have received revelations. The ethic is then bolstered by authority, the authority of the individual philosopher or of the presumed Inspirer. Secondhand acceptance of revelation or of philosophical introspection is usually as intuitive as were the original introspections. The acceptability of an ethical system has generally corresponded with wishful thinking in the individual, who firmly believes what he enjoys believing, and with some sort of pragmatic validation in society. Conviction that the system is right is confirmed by the feeling that it works. The individual is comforted and made happier by a system that responds to his own wishes (so obscure in origin), and any particular society does operate better by erection of standards automatically, although seldom perfectly, adjusted to that society by the fact that they arise and evolve in it.

The mass of people usually find that their own introspective judgment of right and wrong, the edicts of the authorities accepted by them, and the conventions of their society coincide rather closely. They coincide because their sources are related and because the individuals in society tend to modify them or to ignore their discrepancies so as to produce the illusion, at least, of coincidence. The martyrs who cling to a supposedly revealed ethic that is not accepted by their society and the social rebels whose introspective standards reject convention are relatively few in number—although, of course, new systems which may later become conventional arise among such martyrs and rebels.

In all cases satisfaction with the accepted system of ethics, the feeling of a filled need and of emotional security, has depended in large measure on conviction that the system is absolute. It is believed to provide standards of right and wrong which are eternally and everywhere valid, for which the individual has no responsibility other than simple acceptance.

The basis for belief in the intuitive systems of ethics with their claims of eternal and ubiquitous validity was thoroughly shaken by the discovery that they are, in fact, highly relative. Their validation by intuition and their introspective origin itself were found by the psychologists, largely as a result of the Freudian revolution, to depend largely on learning and psychological conditioning, especially in early childhood. They are relative to these processes and their intuitive validation is equally strong in ethical systems diametrically opposite in tendency. The anthropologists found further that extremely diverse ethical systems exist in different societies, all equally valid as far as pragmatic or other objective tests go, and that they are decidedly relative to social structure and other factors that have nothing to do with their absolute validity.[2]

This demonstration that intuitive ethics are relative and cannot, by their very nature, provide an absolute ethical criterion spurred a search that had, indeed, become widespread even before the demonstration was complete. The attempt was to reach an ethic by a process as little intuitive

2. The classic demonstration of these facts is in E. Westermarck's *Ethical Relativity* (London, Paul, Trench, Trubner, 1932).

as possible, to turn to nature, the material cosmos, and to try to deduce standards of conduct from its objective phenomena. The search for a naturalistic ethic was begun long before Darwin, but it was prematurely intensified in the latter part of the nineteenth century with recognition that evolution is a phenomenon of nature which must have ethical bearings. The intensification was premature because the evolutionary process was still insufficiently understood and its possible ethical significance similarly subject to misunderstanding.[3]

The search for a naturalistic ethical system has had as its basis confidence in observation and experiment as leading to discovery of objective truth, or increasing approximation to it, and the conviction that what is ethically right is related in some way to what is materially true. On the negative side, the premise is that there is no necessary relationship between the results of introspection alone and truth or right. These propositions are themselves subjective value judgments and their validity is open to discussion, which it has already received in large measure. A good case can be made out in detail for these convictions, but perhaps their best justification lies in the fact that, even if they had to be taken as articles of faith, they would still be in large measure self-justifying in being fully consistent with our knowledge

3. These earlier attempts to find a naturalistic ethic were reviewed by W. F. Quillian, Jr., *The Moral Theory of Evolutionary Naturalism* (New Haven, Yale University Press, 1945). Quillian seems to have been unaware that, when he wrote, there were few students of evolution who held the views he ascribed to them. Julian Huxley has well reviewed and refuted Quillian's book in his introduction to *Touchstone for Ethics,* already cited. Quillian himself has modified the intensity of his former hostile attitude (personal communication), but his book continues to be valuable as a historical review of earlier developments in the field of naturalistic ethics.

of nature and of man's place in nature. This seems to me to be sufficiently evident in preceding and following discussion and at this point I am inclined to abbreviate by simply postulating these principles not as ethics in themselves but as the basis on which a valid ethic must rest. Of course this does not involve endorsement of any particular ethical system erected ostensibly on the same basis, for it will quickly be seen that a majority of these will not stand up when judged on this very basis.

Many of Darwin's immediate followers thought of evolution primarily in terms of struggle, a concept consistent with and based on Darwinian selection, but not carried to such excess by Darwin himself. They evolved what T. H. Huxley called "the gladiatorial theory of existence" and concluded that the evolutionary ethic must be, first, every man for himself, then every tribe, every nation, every class, and so on, for itself in the "struggle for existence." Such tooth-and-claw ethics suited the book of Victorian laissez faire capitalism and also, with only rather superficial remodeling, of some opposing socialistic ideologies. Those who were not suited by this wholly superficial interpretation of an ill-understood natural process found themselves in a serious dilemma. They could not believe that unbounded personal competition, exploitation and dominance of one group by another, class and national warfare, and the other concomitants of a gladiatorial existence were ethically right. But on the other hand they believed that these are the essential features of the evolutionary process. T. H. Huxley concluded that evolution, although it is a fact that must be faced, is ethically bad. Man's problem then becomes not the forwarding of the evolutionary process but its thwarting. Huxley accepted "the essential evil of the world" and only hoped it could be abated by ethical human conduct. He endorsed and accepted the intuitive ethics, of his time as non-, in fact anti-, evolutionary and did not even discuss their origin or validity.[4]

Some repercussions of this unreal dilemma and of the fallacious tooth-and-claw ethics are still sadly with us. Their influence has been unqualifiedly pernicious. The fact that this once widely hailed product of the search was certainly bad has cast a shadow over the whole effort to find a naturalistic ethic, and should certainly make us extremely cautious in our continuance of that search.

The tooth-and-claw ethic was based on the propositions that evolution as a whole has been ethically good and that its process is gladiatorial. T. H. Huxley saw no alternative to tooth-and-claw ethics except the opposite conclusion that evolution as a whole is ethically bad. The dilemma thus set up was unreal on both sides, for it was neither established that the tooth-and-claw ethic is inherent in evolution as a whole nor that what is inherent in evolution as a whole constitutes a fit guide for human life. Aside from details, basic misapprehensions regarding evolution, as we now know it, were involved. Literal struggle is *not* the essence of natural selection, and natural selection itself is only one of many different factors in evolution. Attention was focused on speciation and the process of speciation was falsely viewed as a competitive struggle between groups of organisms. Furthermore other far more important aspects of evolution were wholly overlooked and the whole course of history viewed in terms of speciation only. There was no reason why speciation, even if it had been correctly understood, should have been taken as *the* basis for an evolutionary ethic. The tooth-and-claw ethic was thoroughly unjustified. Its

4. See his Romanes Lecture, reprinted by Julian Huxley in *Touchstone for Ethics.*

A similar idea was developed at greater length by Sir Arthur Keith in an extraordinary book, *Evolution and Ethics* (New York, Putnam, 1946), whose date makes it a gross anachronism.

excesses and its truly unethical conclusions do not follow from the principles of evolution, and this failure need not put an end to the search for an evolutionary ethic.

Herbert Spencer, still under the sway of the gladiatorial theory of evolution current in his day, was among the first to seek to avoid the ferocity of the conclusions that seemed to arise from this theory by proposing one of many varieties of what may be called "life ethics." The postulate is that life is good. Ethical conduct is that which promotes life. Evolution is good, on the whole, because it has promoted life, also on the whole. From determining how evolution has best accomplished this, we can draw guidance as to how we too can promote life. If our conduct does promote life, it will be ethical. Evolution is not over-all a safe guide, but must be viewed selectively. Extinction is a normal part of evolution, but it does not promote life and is therefore ethically wrong. These are some of the conclusions characteristic of the different sorts of life ethics, of which many variants have been upheld by students of evolution.

Baldly put, the life ethic becomes simply a survival ethic: what best promotes survival is good, what endangers survival is bad. In logical extreme, this becomes only another variant of the old tooth-and-claw ethic. If individual survival is the ultimate good, then it is every man for himself again. (And incidentally, the man who dies for a cause is bad, whether the cause be bad or good, the father who endangers his own life in attempt to rescue his child is bad, and so on.) This extreme is avoided by making group or racial, not individual, survival the goal, but still the ethic is not satisfactory.[5] Of course survival is a condition for any

5. Quillian's attack on naturalistic ethics, previously noted, made a telling point in emphasizing that this particular variety of evolutionary ethics is based on implicit belief in an unconditional obligation to survive, or to avoid extinction. The obligation is not demonstrated, nor is it clearly good or worthy.

progress or for any continuing medium in which good can exist, but what is basically good, in a moral sense, about survival in itself and what part of the evolutionary process could lead us to take this as an ethical guide? Survival is well assured, we have seen, by arrested evolution, by equilibrium in a static environment. Man himself, who needs the ethic, would never have arisen by such survival. Survival for a longer or shorter time is gained by the most diverse and contradictory means, by progression and by retrogression, by specialization and by adaptability, by parasitism—by such a variety of means that to call them all "good" becomes meaningless and certainly offers no hope of an ethical standard for man. And do not forget that survival is ultimately impossible. All life will cease some day. We may and do hope, surely, that mankind will survive as long as possible, but survival will have no ethical significance except as man be good or bad by other standards than merely existing. Certainly it is "good" to survive, but we fall into a semantic trap if we think that this is an *ethical* good.

By a logical extension, the survival ethic becomes an ethic of satisfaction and harmony. It is "good" to survive, for the individual, for the species, and ultimately for all of life. As Leake has said in a development of this thesis, "Relationships between the individuals and groups in contact with each other to be 'good' must therefore be conducive toward the survival of all concerned." Then the general principle appears that "The probability of survival of individual, or groups of, living things increases with the degree with which they harmoniously adjust themselves to each other and their environment."[6] This is the very opposite of the tooth-and-claw ethic. It is based on a much truer conception of the

6. Both quotations are from the following thoughtful and valuable paper: C. D. Leake, "Ethicogenesis" (*Sci. Monthly, 60* [1945], 245–253). See also C. D. Leake and P. Romanell, *Can We Agree? A Scientist and Philosopher Argue about Ethics* (Austin, Univ. of Texas Press, 1950).

evolutionary process and of the processes of life in general. It also has the virtue of being congruous with almost any ethic, except the plainly false tooth-and-claw ethic, that can be based on the evolutionary process. That is, almost any other naturalistic system of ethics would agree that behavior consistent with this principle would usually be ethical.

Action on this principle would usually tend to *promote* good, and yet it does not seem to be a basic criterion for judging what *is* good. It still departs from the postulate that survival is "good," and we have noted that this is not necessarily true in an ethical sense. Moreover, this is a purely static concept. It would seem logically to involve a changeless equilibrium, with maximum or complete harmony, as the highest good. If this implication is accepted, then the principle is unrealistic and is not evolutionary, nor surely good. The tooth-and-claw ethic is invalid, but there are teeth and claws in nature. Some organisms must die if others are to live. And evolution involves ceaseless change. Even extinction of races and of whole orders and classes has been a condition of the rise of higher types. The harmonious equilibrium is never reached, and if it had been, man would not now exist. The equilibrium is not achieved with the coming of man, nor is there apparent opportunity for him to attain it.

A further step in the search accepts and emphasizes the dynamic nature of life and of its evolution. Departing, still, from the proposition that life is good, the dynamic inference is that increase in life is good. Whatever tends to increase life is therefore right, and whatever tends to decrease it is wrong—an ethic of increase or of abundance of life.[7] Here again, action under this principle would sometimes be ethical

7. This idea has often been expressed in various forms. A well-reasoned version of it was given in a previous volume of Terry Lectures: H. S. Jennings, *The Universe and Life* (New Haven, Yale University Press, 1933).

by most other standards of ethics, but reflection raises doubts as to whether the principle is itself ethical, that is, a standard by which right and wrong action may be judged. Its applicability is narrower than that of the ethic of harmony, which does not contradict it. How many concrete situations can you think of in which you could decide on right and wrong action, as an individual or for a nation or for mankind as a whole, on the basis of consequent increase or decrease in life? Murder becomes bad, to be sure, but most situations which do have ethical bearings elude application of the principle. Other problems, where it does somehow bear, require some other and overriding principle for solution. It is good to plant a crop, but is it good or bad to harvest it? And is this situation, to which the principle definitely applies although the direction of the application is equivocal, really an ethical one at all? Evidently if the principle is ethical it is severely limited and inadequate as basis for a system of ethics.

Further thought must raise doubt as to whether the principle would indeed lead to right and to the avoidance of wrong. Which is right and which wrong: an earth so crowded with men that they can barely subsist by laboring all their waking hours and any further increase is cut off by starvation, or an earth moderately populated and with ample provision for all? The alternatives now exist as between, say, parts of China or India and most of the United States. Trend toward maximum population and minimum subsistence is necessarily right and good under the increase ethic. Judgment that it is wrong and bad would be suspect if merely intuitive, but as will appear, this judgment does follow from an ethic more soundly based in the history of life than is the increase ethic.

Various forms of the ethic of increase of life are commonly derived from or supported by the statement that this constitutes a major trend of evolution. This ethic then becomes one of many, sometimes contradictory, possible and proposed ethical systems which depend on the postulate that continuation of the trend, or of a trend, in past evolution is an ethical guide to right conduct and to future good. Such a conclusion follows almost automatically from a finalist view of evolution, especially when this is put on a theistic or otherwise religious basis. If evolution has a goal, if it has all been a progression toward some ultimate and determining end, then we should find out that goal by observations of the facts of evolution and it is reasonable to consider further progress toward the goal as ethically right. This becomes not only reasonable but also obligatory if the goal was set by God. Unfortunately for hopes for so simple a solution of the ethical problem, the basis for this is flatly nonexistent. There is no real evidence whatever that evolution has had a goal, and there is overwhelming evidence that it has not.

Evolution does have trends, many of them, and it has been seen that the most nearly general of these has been toward increase in the total of life on the earth. Hence the increase ethic would seem to be validated if trend ethics were acceptable as valid. But all trend ethics demand the postulate that the trends of evolution or some particular one among these is ethically right and good. There is no evident reason why such a postulate should be accepted. It is, at least, impossible to conclude that *all* evolutionary trends are ipso facto good. They are too diverse and they lead as often to decrease of life and to ultimate extinction as to increase and survival. The problem arises that a choice must be made; it is necessary to decide *which* trends are good. The decision cannot, then, be based on factual observation that a trend has occurred but must involve some other and quite distinct ethical criterion. Any attempt to find a valid trend ethic results merely in a requirement that some other ethic first be found in order to select the right ethical trend. Thus the increase ethic is not validated because it follows a general trend in evolution, but on the contrary, identification of that trend as right depends on prior admission of the validity of the increase ethic—an admission which, as previously shown, is neither forced nor warranted by what is known of the evolutionary process.

There is another group of attempts to derive ethics from aspects of the evolutionary history of life which are often, but not necessarily, supported by the fallacious argument from trend. These proposed systems, which might collectively be called aggregation ethics, see as ethically good the increased aggregation of organic units into higher levels of organization. One form of the argument runs more or less as follows: Evolution has involved a succession of organic levels, with progressive complication and perfection of coordinated structure and function on each level. The protozoans and other one-celled (or noncellular, or cellularly undivided) organisms represent a lowest level. The next level is that of multicellular individuals, metazoans, with increasing differentiation of the cells and their grouping into organs with increasingly specialized functions. There follows another level, the highest, in which metazoan individuals were aggregated into hyperzoan organisms. The hyperzoan or "epi-organism" is society, into which individuals are to merge and be integrated as subordinate parts of a higher whole. It is, in fact, the so-called organic state, considered as having an individuality and life of its own. In such a state individuals, or "persons," as some adherents of this view prefer to call them, exist for the state, the good and rights of which are separate from and superior to that of the individuals composing it.[8]

8. The organismic theory of the state was pre-evolutionary and in its modern form seems to stem directly from Comte. It was left for some later biologists to "discover," wholly falsely, as I believe, that the trend of evolution confirms the organic nature of the state, makes this biologically "right," and establishes as the basis for ethics the promotion of the state as an organism. The whole complicated subject of organic levels

The whole argument and its ethical implications among which was support of authoritarian and totalitarian ideologies as biologically and ethically right, were thoroughly erroneous. When the state or any other social structure is called an "organism," the word is being used in a way fundamentally different from its use for a biological organism such as an ameba, a tree, or you. The state is not an individual or a person in anything like the same sense that these organisms are individuals. The parts that compose it retain, and indeed intensify, their organic individuality. Their relationship to the social unit is entirely different from the relationship of cells or organs to a metazoan individual. Calling a state an "organism" and concluding that it is therefore comparable with a metazoan organism is a glaring example of the fallacy of the shifting middle term. Use of the comparison as an analogy provides an interesting descriptive metaphor, but its use in support of an aggregation ethic is a particularly egregious misuse of analogy, confusing it with equivalence and extending it as an interpretive principle far beyond the point to which it is valid even as metaphorical description.

Furthermore, if the fallacious use of "organism" as a shifting term is avoided, it is quite evident that merging of the individual into a higher organic unit is not a common trend in evolution and, specifically, is not at all a trend in human evolution. The trend in human evolution and in many other evolutionary sequences has been, on the contrary, toward greater individualization. The particular type of social orga-

---

was rather fully treated from many points of view, including that of the "epi-organism school," in R. Redfield, *Levels of Integration in Biological and Social Systems* (Lancaster, Pennsylvania, Jaques Cattell, 1942). After the horrid examples of Nazism, Fascism, and Hitler's War, I believe that few if any western biologists continue to push the concept of the organic state as a guide to ethics.

nization characteristic of man, as opposed, say, to that in an anthill, has been based on high individualization of its members and has intensified this. It will be necessary to refer again to this fact, and to clarify some of its implications, in the next chapter in connection with what I believe to be a truer conception of the ethical significance of man's place in nature and of his evolution.

These various attempts to find a naturalistic ethic where it would seem most likely to reside, in the process of evolution, have certainly clarified the issues involved. They have narrowed the field of search and most of them have, at worst, skirted an acceptable solution. At best they have produced partial answers that are indeed ethically good although not achieving a general and firmly based evolutionary ethic. It is now clear that none of the attempts reviewed up to this point can be considered satisfactory. All were seeking a more coherent and realistic basis for ethics than could be found in the older intuitive systems, and in this we must conclude that they were on the right track, but in so doing they have not escaped all the pitfalls of those older systems.

Most of these naturalistic ethics have been attempts, as were intuitive ethics, to find standards of right and wrong with eternal validity for all things from ameba to man. This adherence to an ideal absolutism fails in appreciation of two basic facts of evolution: that change is of its essence and that man represents a new, unique sort of organism subject to a new, distinct sort of evolution in addition to the older sort, which still continues. The need and the search for an ethic are also unique to man and in fact ethics enters into evolution only at this point; ethical judgment arises in the new evolution and is one of its characteristics. There are no ethics but human ethics, and a search that ignores the necessity that ethics be human, relative to man, is bound to fail. Attempts to derive an ethic from evolution as a whole, without particular reference to man, are further examples of the "nothing but" fallacy of the nature of man.

Another basic weakness in not quite all but most of these attempts has been that they involve trying to find out what evolution has been up to, or even what evolution seems to *want,* and then assuming that promotion of this is ethically right. This amounts to placing evolution in the position occupied by God or by His revelation in intuitive ethics. Such naturalistic ethics share with those they attempted to replace a certain evasion of responsibility. They still try to find an external standard, one given without need for choice and without other requirements than discovery and acceptance.

So the enquiry seems to have swung in a circle and to be back rather near where it started. Yet there is a point from which to go on, and the unsatisfactory nature of these attempts does not preclude the possibility of others that can be satisfactory. The point is that an evolutionary ethic *for man* (which is of course the one we, as men, seek, if not the only possible kind) should be based on man's own nature, on his evolutionary position and significance. It cannot be expected to arise *automatically* from the principles of evolution in general, nor yet, indeed, from those of human evolution in particular. It cannot be expected to be absolute, but must be subject to evolution itself and must be the result of responsible and rational *choice* in the full light of such knowledge of man and of life as we have.[9]

9. I have recently discussed the subjects of this and the following chapter from a different but not contradictory point of view in "Naturalistic Ethics and the Social Sciences" (*Amer. Psychologist, 21* [1966], 27–36).

XIX. THE ETHICS OF KNOWLEDGE AND OF RESPONSIBILITY

The meaning that we are seeking in evolution is its meaning to us, to man. The ethics of evolution must be human ethics. It is one of the many unique qualities of man, the new sort of animal, that he is the only ethical animal. The ethical need and its fulfillment are also products of evolution, but they have been produced in man alone.

Man's knowledge that he exists is, at the least, more conscious and particular than that of any other animal. Man alone knows that he has evolved and is still doing so. Man alone places himself in a conceptual framework of space and time. Man possesses purpose and exercises deliberate choice to a unique degree, even if, indeed, these capacities can be said to be the same in kind in any other animals. Man foresees the results of his actions. It is most improbable that any other animal has more than an inchoate or largely instinctual sense of values, while in man this is normally conscious, orderly, and controlled. (This does not contradict the fact that, even in man, the *origin* of his valuations is in considerable part unconscious and may be quite uncontrolled.)

Conscious knowledge, purpose, choice, foresight, and values carry as an inevitable corollary responsibility. Capacity for knowledge involves responsibility for finding out the truth and, in our social system, for communicating this. Foresight makes possible choice of action by evaluation of its probable result. The possibility of choice brings an ethical responsibility for selection of what is right. The sense of values implies means and responsibility for decision as to what is right. Purpose confers the power and, again, the responsibility for translating choice and value into right action. These capacities and responsibilities are not qualities of life in general or of its evolution, but specifically of man. Man is much the most knowing or thinking animal, as our predecessors rightly recognized in bestowing on him the distinctive qualification of *sapiens.* Man is also the responsible animal. This is more basic than his knowledge, although dependent on it, for some other animals surely know and think in a way not completely inhuman, but no other animal can truly be said to be responsible in anything like the same sense in which man is responsible.

The search for an absolute ethic, either intuitive or naturalistic, has been a failure. Survival, harmony, increase of life, integration of organic or social aggregations, or other such suggested ethical standards are characteristics which may be present in varying degrees, or absent, in organic evolution but they are not really ethical principles independent and absolute. They become ethical principles only if man chooses to make them such. Man cannot evade the responsibility of the choice. As his knowledge embraces facts about these characteristics in evolution, they become part of the basis on which his ethical principles should be developed, but they supply no automatic guide to good and bad.

Man has risen, not fallen. He can choose to develop his capacities as the highest animal and to try to rise still farther, or he can choose otherwise. The choice is his responsibility, and his alone. There is no automatism that will carry him upward without choice or effort and there is no trend solely in the right direction. Evolution has no purpose; man must supply this for himself. The means to gaining right ends involve both organic evolution and cultural evolution, but human choice as to what *are* the right ends must be based on human evolution. It is futile to search for an absolute ethical criterion retroactively in what occurred before ethics themselves evolved. The best human ethical standard must be relative and particular to man and is to be sought rather in the new evolution, peculiar to man, than in the old, universal to all organisms. The old evolution was and is essentially amoral. One overriding consideration can indeed be derived from the processes of organic evolution: the good of a species is served by such characteristics and changes as are adaptive. The "good" of the old, the organic evolution is not intrinsically ethical. That "good" still applies to the human species, but here it comes to involve also the new evolution and conscious knowledge, including the knowledge of ethical good and evil.

The most essential material factor in the new evolution seems to be just this: knowledge, together, necessarily, with its spread and inheritance. As a first proposition of evolutionary ethics derived from specifically human evolution, it is submitted that promotion of knowledge is essentially good. This is a basic material ethic. "Promotion" involves both the acquisition of new truths or of closer approximations to truth (metaphorically the mutations of the new evolution) and also its spread by communication to others and by their acceptance and learning of it (metaphorically its heredity). This ethic of knowledge is not complete and independent. In itself knowledge is good, but it is effective only to the degree that it does spread in a population, and its results may then be turned by human choice and responsible action for either good or evil.

These considerations suggest some notice of the mechanisms for acquiring knowledge, of values in knowledge, and of its right and wrong utilization—subjects obviously too vast for adequate discussion here and yet too important to be passed over in total silence. Knowledge is of many sorts and is to be sought in no one way. Perception of tenderness and of security acquired at the mother's breast is as truly knowledge as is determination of the fabric of an alloy acquired at the electron microscope. Science is, however, our most successful and systematic means of acquiring knowledge and, at present, almost alone in the power to acquire knowledge wholly new to man. It used to be usual to claim that value judgments have no part in science, but we are coming more and more to perceive how false this was. Science is essentially interwoven with such judgments. The very existence of science demands the value judgment and essential ethic that knowledge is good. The additional and still more fundamental ethic of responsibility makes scientists individually responsible for evaluating the knowledge that they acquire, for transmitting it as may be right, and for its ultimate utilization for good.

A broad classification of the sciences into physical, biological, and social corresponds with three levels of organization of matter and energy, and not levels only, but also quite distinct kinds of organization. The three are of sharply increasing orders of complexity and each includes the lower grades. Vital organization is more intricate than physical organization and it is added to and does not replace physical organization, which is also fully involved in vital organization. Social organization retains and sums the complexities of both of these and adds its own still greater complexities.

Many of the dangers and ethical problems of science arise from these relationships. At one end, the groundwork of scientific knowledge is in the physical sciences, and they can produce far-reaching results, for good or evil. The physical sciences are the simplest of the three groups. Discoveries here are made more readily and more rapidly, so that advances of knowledge in this field are accelerated. But the physical sciences, which for these reasons need for their guidance the most rigid and continuous application of ethical standards, are farthest removed from the source of those standards and the means of their application. At the other end, this source and these means lie in the social sciences, but they are by far the most difficult and intricate. Their discoveries demand much more travail and are slower to be reached, far slower to be disseminated and utilized. Their ethical guidance is no less important than that of the physical sciences, but they have the added complication that they are themselves the lumber room of ethics, crowded with old, conflicting, and certainly partly false ethical systems and the battleground of these. To mention only one dilemma arising from this situation, the physical sciences have put man in possession of the awful secret of atomic energy before the social sciences have produced adequate means of controlling this or of securing its ethical application. The inequity of knowledge is in itself unethical and is one of man's great blunders. It could be his last.

The possibilities of wrong in the unequal development or in the results of the spread of knowledge reveal the need for another ethic, one that may be more profound or, at least, that shall have a moral rather than a mainly material basis. A second essential feature arising from the unique status of man in the history of life was seen to be his possession of personal responsibility. It is now submitted that the highest and most essential moral ethical standards are involved in the fact of man's personal responsibility.

This responsibility is not in itself an ethic. It is a fact, a fundamental and peculiar characteristic of the human species established by his evolution and evident in his relationship to the rest of the cosmos. Recognition of this responsibility and its proper exercise are the firm basis on which right and moral human action must be based. From it arises a pervasive ethic which, among other things, may ensure the right interpretation and action of the ethic of knowledge. Human responsibility requires, in each individual as well as in society as a whole, that the search for knowledge be a search for truth, as unbiased as is possible to human beings; that probable truths as discovered be tested by every means that can be devised, that these truths be communicated in such a way as is most likely to ensure their right utilization and incorporation into the general body of human knowledge, and that those who should receive this knowledge seek it, share in its communication, and in their turn examine and test with as little prejudice as possible whatever is submitted as truth. This is a large order, indeed, but a necessary one. It involves responsibilities for every living person, and responsibilities that cannot be ethically evaded; that is, their evasion is morally wrong. Among other consequences of this morality, it follows that blind faith (simple acceptance without review of evidence or rational choice between alternatives) is immoral. Such faith is immoral whether it is placed in a theological doctrine, a political platform, or a scientific theory.

Of course this does not mean that every individual must become a theologian before it is moral for him to join a church, a political economist before it is moral to vote, or a research biologist before it is moral to believe in evolution. The field of knowledge is obviously far too vast, and human mentality far too limited, for each of us to grasp all these intricacies. It is a social requirement that there be specialists in each field whose profession it is to examine and to test such truths as pertain to it. It is the moral duty of these specialists to submit their qualifications, the results of their judgment, and such general evidence for it as is essential for its substantiation. When such judgments conflict, as they often do on a given point, the moral duty of the nonspecialist is to choose the judgments of that authority whose qualifications are greatest in the pertinent field and whose submitted evidence is best. The individual remains personally responsible for making the choice, even if he must do so on the basis of the knowledge of others and not his personal knowledge.

Neither this nor any other process can rigidly guarantee making the right choice. It is not to be taken as an appeal to authority in the sense of belief in the absolute validity of any one opinion or of infallibility in science or in religion. The important point is responsibility for using the right *method* of choice. The right method is evaluation of evidence and avoidance of pure intuition and of authoritarian dogma. Recourse to authority, in this context, demands judgment that the accepted opinion is based on rational consideration of known evidence. It is the rejection of revelation or of emotional reaction when knowledge is available. It further rejects the absolute nature of any authority. Choice must be prepared to change if increase in knowledge or its better interpretation occurs.

Beyond its relationship to the ethic of knowledge, the fact of responsibility has still broader ethical bearings. The responsibility is basically personal and becomes social only as it is extended in society among the individuals composing the social unit. It is correlated with another human evolutionary characteristic, that of high individualization. From this relationship arises the ethical judgment that it is good, right, and moral to recognize the integrity and dignity of the individual and to promote the realization or fulfillment of individual capacities. It is bad, wrong, and immoral to fail in such recognition or to impede such fulfillment. This ethic applies first of all to the individual himself and to the integration and development of his own personality. It extends farther to his social group and to all mankind. Negatively, it is wrong to develop one individual at the expense of any other. Positively, it is right to develop all in the greatest degree possible to each within the group as a whole. Individuals vary greatly in other capacities, but integrity and dignity are capable of equal development in all.

Socialization and individualization may conflict, but they may also work together for the advancement of each. Here again choice is possible, and not only a possibility but also an unavoidable responsibility. Individual integration and welfare can be secured at the expense of others, but they may also be achieved, and reach their highest degree, by interaction which promotes others along with the self. Under this system of ethical standards a definition of the good society might indeed be simply that it is a society in which this interaction is usual.

An individual in society leads no existence wholly apart from that society, any more than the most solitary individual of any species of organisms can exist without reference to its environment. The social group is part of the human environment, a largely self-constructed part. It is the medium in which the individual exists and it is one of the molding, evolutionary influences on the individual. The abstraction of a human individual not so molded and influenced is completely unreal. Flexibility in reaction to the environment, especially including the social environment, is a major characteristic of the human species and is the essence of human individualization. A human organism that developed outside of such a framework, if this were possible, would certainly be far less individualized than one developed within it. Human personality and accompanying individualization, with all its amazing variety, depend on interaction with the environment to a far greater degree than does the lesser individualization of any other organisms. The sorts of human society that we call civilized all demand and are based on a rather high minimum of variability and of individualization in its members, a minimum far above that reached by any nonhuman species. In turn, such a society provides the possibility of degrees of individualization far greater than could ever be achieved outside it. In biological terms, it provides a variety of ecological niches much greater than those of primitive tribal society and tremendously greater than those available to any other one species of organisms. The niches are filled by individual adjustability to larger extent than by genetic adjustment.[1]

The apparent conflict of socialization and individualization thus does not really exist, and suggestion of choice between one or the other presents unreal alternatives. Individualization is a means of socialization and socialization provides enriched opportunities for individualization.[2]

Another equally false set of alternatives is presented by any contrasting of collective action or security with personal responsibility. The collective aspects of the state are, or ethically should be, achieved by means of personal responsibility in all its members. Collective social measures, including provision for the underprivileged, are, or ethically should be, undertaken to prevent unethical development of some individuals at the expense of others and to promote the ethical equal development of all to the extent of their capacities. Such measures should not, and in the nature of things cannot, replace or eliminate personal responsibility but on the contrary require this and guarantee it, if they are both ethical and effective.

Collective action to promote individual development and to prevent exploitation is evidently required in an ethically good society, but its results can also be ethically bad. It can reinforce but can also contravene the ethical principles of

1. The plasticity of man and his great capacity for personal adaptation are especially well discussed by Th. Dobzhansky in *Mankind Evolving* (New Haven, Yale Univ. Press, 1962), and by R. Dubos in *Man Adapting* (New Haven, Yale Univ. Press, 1965).

2. A distinct but parallel, equally unreal dichotomy is sometimes presented in the study of organic, rather than social, evolution. Only populations evolve, but it is individuals that do the breeding, the being born, the living, and the dying that produce that evolution. The false alternatives are involved in the query whether it is the population or the individuals that determine the process. In view of the facts of the process, these alternatives simply do not exist and the question is essentially meaningless.

personal responsibility and of individual integrity and dignity. This inevitably results if the members of society feel that their responsibility is delegated in the collective action and ceases to be individual. The individual who needs protection and help is not relieved of his responsibility for himself by the existence of a social mechanism for his assistance. Under our ethic, it is equally wrong for the individual to fail to seek such help as he needs and for him to demand or even to accept such help as a replacement for self-responsibility. On the other side, the social provision under this ethic must, if it is to be ethically right, be only the required means by which each individual carries out his personal responsibility toward any other individual who needs protection or aid. The responsibility is not diminished and it remains in the individual; it is not transferred to the mechanism that implements it.

It is fundamental in all this that responsibility is rooted in the true nature of man. It has arisen from and is inherent in his evolutionary history and status. Responsibility is something that he has just because he is human, and not something that he can choose to accept or to refuse. It cannot be rejected or unconditionally handed over to others. The attempt to do this is ethically wrong, and the responsibility remains where it was. The delegation of responsibility, to the extent that this is possible and proper, involves continuing responsibility for the actions of the delegate. In the last analysis, personal responsibility is nondelegable. Not only is every individual personally responsible for any actions by delegates or representatives of his, but he cannot ethically, even in semblance, delegate any responsibility for his own actions and for all their results.

The sweeping and impersonal nature of these generalizations may suggest difficulties and exceptions when particular and personal situations are considered. Is, for instance, the individual or society responsible if a jobless workman steals to feed his starving family? On reflection, this will be seen to be another false alternative. Everyone involved is necessarily responsible. The workman is certainly responsible for his action. On the stated facts alone the action itself may be ethically either good or bad. It is good if there was no alternative involving a lesser total of injury and indignity, determined by striking a balance between the good done for the family and the harm done morally to the workman himself and materially to his victim and to others who might otherwise have obtained the stolen food or who must indirectly pay for it. It is bad if avoidable harm is done, if, for instance, the workman has merely evaded responsibility for obtaining help in the social services available or for helping himself in a more proper way, such as accepting or seeking work. And in any case all other individuals in his society have a responsibility, which they may or may not have lived up to in the given case, first to see that sufficiently compensated work is available to the workman and second, in case of unavoidable failure in this, to see that the family is helped when help is needed beyond their own capacities.

As one more example, there is the old ethical problem of responsibility when an army officer orders a private to kill another human being. The situation is inherently bad: destruction of an individual is necessarily unethical on our premises. In a truly good society there would be no officer and no private. Pending the arrival of a good society, it is possible for such evil to be overbalanced by good resulting from the action. Whether the action was, in fact, more conducive to evil or to good in the particular circumstances, the officer and the private both have complete personal responsibility for it. The officer could not delegate the responsibility by giving an order and is as responsible as if the shot were fired by his own hand. The private who does fire the shot cannot turn responsibility for his personal action back on the person or the circumstances inducing him to take the action.

He has acquiesced in a situation where he knows that he is bound to carry out the orders of others. It was right for him to do so if in his judgment the ultimate result would be better than the result of his refusal to acquiesce. He remains responsible not only for this decision but also for any immediate actions, good or bad, that result from his being under orders.

Ethical standards based on the fundamental evolutionary characteristics of man, particularly on human knowledge and human responsibility, have the widest applications to human conduct and respond fully to the human need for judgment of right and wrong. The broadest problem now facing mankind is choice between conflicting ideologies. These evolutionary ethics here lead to unequivocal decisions.

Authoritarianism is wrong. The assignment of authority on a fixed basis, without constant check and periodic review, inevitably involves an attempt to delegate nondelegable responsibility and to evade responsibility for subsequent actions of the delegate. This is an ethically wrong denial of the personal responsibility inherent in man's nature. The system inevitably leads to exploitation of all by the authoritarian leader and the development of a hierarchy in which each higher group exploits those below it. This is a morally wrong development of some individuals not along with but at the expense of others.

Totalitarianism is wrong. The concept of a state as a separate entity with its own rights and responsibilities contravenes the biological and social fact that all rights and responsibilities are vested by nature in the individuals that compose the state. The claim that the welfare of the state is superior to that of any or all its component individuals is thus absurd on the face of it, and the claim inevitably is used to excuse denial of maximum opportunities for individual development by all, a denial which is immoral. Concomitants sure to occur in such a state are regimentation of individuals, policing of all conduct (including conduct ethically good), judgment of knowledge and of ideas on a basis other than that of their truth and falsity, and consequent suppression of truth and unethical control of the dissemination of knowledge. Totalitarianism cannot conceivably be reconciled with any ethical system that admits the goodness of knowledge, the ineradicable existence of personal responsibility, and the value of individual integrity and dignity.

Democracy is wrong in many of its current aspects and under some current definitions, but democracy is the only political ideology which can be made to embrace an ethically good society by the standards of ethics here maintained. Laissez faire capitalism, or any other societal activity that promotes or permits selfish or unfair utilization of some individuals by others, is obviously wrong by these standards. Capitalism, not further restricted, is perfectly consistent with authoritarianism or totalitarianism and is of course wrong if involved in either of those morally wrong systems. In a socialized democracy, controlled capitalism without improper exploitation may be ethically good. Majority rule is wrong if it involves suppression or oppression of any minority, but decision of problems by the majority of all those affected by them, accompanied by free expression of all opinions and full preserval of minority rights is, so far as has yet been demonstrated, the only possible ethically good means of reaching collective action. Attempts to assign personal responsibility to the government are ethically wrong (and biologically futile, to boot), but government by representatives or delegates for whose actions each one of their constituents remains personally responsible seems to be the only practicable method of ethical government for large groups of people.

Governments called democracies are by no means all ethically right by our standards, and none is free of many ethically bad aspects. Yet an ethically good state, one based on the fact of personal responsibility by each of its members and organized to promote the acquisition, dissemination, and acceptance of truth in all fields, to maintain the integrity and dignity of every individual, and to enable maximum possible realization of personal capacities—such a government would necessarily be a democracy.

Below these broadest aspects of the ethical problems of our times, application of these evolutionary ethics to personal problems and actions may be complex but is usually pertinent and decisive. A basic problem for all of us is the integration and ethical development of our own personalities. A major factor in this is certainly the acceptance and exercise of our personal responsibility. There is often no essential clinical difference in neurotic trends and strains or underlying personality factors between those who are socially normal, workably adjusted and integrated and those who are pathologically maladjusted.[3] The difference is essentially that the former have the ethically good will toward integration and take the responsibility for achievement of this in themselves. It follows, of course, that the maladjusted are in such cases (which admit of certain exceptions, as in the case of some traumatic psychoses) responsible for their own condition. The responsibility extends to the necessity for seeking help, from psychiatrists or others, when personal resources are not wholly sufficient, and it is a responsibility of every other member of society to see that such help is available and sufficient for those who do seek it.

It is, indeed, another inescapable biological fact that no individual is fully self-sufficient, and the doctrine of personal responsibility carries no such illogical corollary. It is a

3. It is important to insist on this point. It refutes with sufficient evidence from psychology itself an erroneous impression that post-Freudian views on the development of personality and of ethical mechanisms necessarily dilute or eliminate self-responsibility. See A. Roe, "Man's Forgotten Weapon" (*Amer. Psychologist, 14* [1959], 261–266).

responsibility not only with relationship to the self but also with relationship to others, ultimately embracing all of mankind, a responsibility for cooperation and for both giving and obtaining aid. This aspect of evolutionary ethics, above all others, begins in the cradle and follows us to the grave. Normal development of the human infant requires the presence of a mother and frequent contact with her.[4] Her responsibility in this matter is another that cannot be delegated. Withholding of continual affectionate contact or unnecessary assignment of care to an impersonal agency is immoral because it impedes normal integration of the infant's individuality.

In such a situation, and in many others, the bearing of these ethics is perfectly clear. They provide, however, no guide by which every action of daily life need be judged and channeled into a single possible direction. In most respects wide latitude remains between "right" and "wrong." It is, in fact, a part of the sense of values arising from these considerations that diversity in personality and in action is a good and positively valuable characteristic of mankind, so long as it does not overstep broad ethical bounds or impinge unfavorably on other personalities. Regimentation of personality or suppression of individuality is ethically bad by these standards. Respect for others is ethically good. Appreciation rather than bigoted disapproval of differences from ourselves is one of the means of enriching our own experiences and personalities, and this too is one of the ethical goods.

It should, finally, again be emphasized that these ethical standards are relative, not absolute. They are relative to

4. "Mother" in this sense cannot always be the biological mother and the process may be sufficiently normal with a mother substitute. It is, however, better for the mother to be the biological mother. The infant does not know the difference, but the mother herself does and this affects the psychological situation.

man as he now exists on the earth. They are based on man's place in nature, his evolution, and the evolution of life, but they do not arise automatically from these facts or stand as an inevitable and eternal guide for human—or any other—existence. Part of their basis is man's power of choice and they, too, are subject to choice, to selection or rejection in accordance with their own principles. They are also subject to future change as man evolves; after all, if mankind does pursue the ethic of knowledge it should be able progressively to improve and refine any ethical system based on knowledge.

An ethical system is a necessary part of the specific adaptation of *Homo sapiens* to his environment, which of course for each individual includes his fellow men. Environments change, and this environment is changing at an altogether exceptional rate. It follows that maintenance of adaptation may or, in all probability, will require modification of adaptive characteristics including ethics. But in cultural evolution as in biological evolution, change cannot occur unless there is variation. It is therefore desirable and necessary for future progress that there be variation in ethical codes among human populations. It is, however, also essential that all variants be viable, adaptive to at least a minimum degree, not definitely antisocial under conditions actually existing.

There is no ethical absolute that does not arise from error and illusion. These relativistic ethics have, at least, the merit of being honestly derived from what seems to be demonstrably true and clear.

XX. THE FUTURE OF MAN AND OF LIFE

Evolution is clearly going on around us today. There is no obvious reason for its not continuing for many millions of years to come, as regards both man and the rest of the organic realm. It would be futile, in the present state of knowledge, to try to predict just what this future evolution will produce. It is, however, an essential part of study of the meaning of evolution to try to assess its present status and, not the future course, but the possibilities for the future inherent in the present condition of man and of other forms of life. Such assessment, made periodically as our knowledge increases, is also a necessary basis if we are eventually to make use of the unique human power to exercise a measure of control over evolution.

Some authorities[1] have, indeed, maintained that evolution is through, that life has evolved as far as possible and all essential evolutionary change stops at this point. A finalist could perhaps believe that evolution had a single aim, such as the production of man, and that it has stopped with achievement of that aim. But evolution is not, in fact, final-

1. Notably R. Broom, for instance, in "Evolution as the Palaeontologist Sees It" (*South African Jour. Sci., 29* [1933], 54–71). Broom often returned to this point in later discussions of his extraordinary finalistic and metaphysical theory of evolution. Although I know of no one who now takes Broom's general theory seriously, a number of other students of evolution agree with him on this particular point, among them Sir Julian Huxley, who, however, makes an exception for man. He has expressed the opinion that future evolution will occur *only* in man.

istic and it seems to me quite impossible for it to come to a standstill as long as there are any living organisms left to evolve. Life and its environment are in such ceaseless flux that it is simply inconceivable that a permanent equilibrium will ever be reached. Inability to predict just what evolution will do next is a sign of our ignorance, not of the impotence of evolution to do something more.

If an intelligent being had been able to study the world of, say, the Jurassic or Cretaceous without knowledge of what was to follow, he would have had as much reason as we have now to think that evolution had exhausted all possibilities. The barrel of life must then have seemed full to overflowing, with no opportunities overlooked. In reality, we know that extensive replacement has occurred in the ways of life then already occupied, and that new ways have in fact also been developed. It must also have seemed then that all existing organisms were too specialized to give rise to any important new radiation. The few small and obscure mammals would have appeared to be merely highly and peculiarly specialized reptiles. It would seem extremely rash to predict that no form of life now in existence can possibly have similar potentialities for radiation on a new and unforeseen basis.

It has been suggested that *all* animals are now specialized and that the generalized forms on which major evolutionary developments depend are absent. In fact all animals have always been more or less specialized and a really generalized living form is merely a myth or an abstraction. It happens that there are still in existence some of the less specialized, that is, less narrowly adapted and more adaptable, forms from which radiations have occurred and could as far as we can see occur again. Opossums are not notably more specialized now than in the Cretaceous and could almost certainly radiate again markedly if available spaces were to occur again. Even primate re-radiation would be entirely possible if present primates were wiped out and their empty ecological niches continued to exist, because living tree shrews have no specializations that would clearly exclude approximate repetition of such a radiation. Of course such re-radiation could not be identical with or even closely similar to the radiations that have already occurred in the same groups, because close repetition of the same conditions is impossible. It is also extremely improbable that re-radiation will occur, simply because the results of earlier radiation are there, holding their ground; but the possibility certainly is not excluded. More likely, in view of the past record, would be either replacement or radiation of quite a new and unpredictable sort—unpredictable because we probably do not know enough to see the possibility. (Such possibilities can be imagined, although hardly predicted; for instance there are no truly aerial flora and fauna of organisms living and reproducing in air as a medium as seaweeds and fishes do in water, and there seems to be nothing downright impossible in such a development.)

If man were wiped out, it is extremely improbable that anything very similar to him would ever again evolve, although I cannot see that even this is altogether impossible. The exact ancestral forms are gone. The whole intricate sequence of biological and physical conditions that gave rise to man certainly will not be repeated with very close approximation. Yet it remains true that manlike intelligence and individual adaptability have high selective value in evolution and that other animals have a conceivable basis for similar development. One, at least, of its most ancient and fundamental requisites is actually quite common among mammals, even outside of the primates: use of the forefeet as hands for manipulation, with close hand-eye coordination, which is present not only in all primates but also in many rodents and some carnivores.

It is, however, reasonably safe to assume that no animals

able to compete with man in intelligence, socialization, and the other unique human characteristics will arise as long as man does exist in fact. He has a firm grip on this adaptive zone and is fully able to defend it. The existence of man is a major element in defining and limiting possibilities of future evolution not only in this but also in practically all other respects. Man does broadly manipulate the environment and is learning how to do so more and more. He knows that evolution occurs and is fast learning exactly how it works. This must, if it continues, eventually make it possible for him to guide not only his own evolution but also that of any other organisms, if he so chooses. It is a decided possibility that he can really introduce finalism into organic evolution, which has conspicuously lacked a true goal in the past; the purpose and the end would, of course, be set and determined by man. He is rapidly coming to hold the power of life and death. He has casually caused the extinction of numerous other sorts of organisms and seems likely to devise means for causing extinction at will.

This awesome power includes the human prerogative of self-extinction. It is highly improbable that any organisms have ever become extinct as a result of their own activities alone and without some affliction unbidden and not determined wholly by their own natures. Man is probably quite capable of wiping himself out, or if he has not quite achieved the possibility as yet, he is making rapid progress in that direction.

If man does not exercise this, another of his unique capacities, or if in any case his extinction (ultimately inevitable) is long delayed, it is reasonably certain that he will evolve farther and will change more or less radically. It is hardly conceivable that even man's great powers will ever include the possibility of maintaining a frozen status quo in a changing universe. At present, in any case, cessation of human evolution is certainly not desirable or desired. By no standard of ethics whatever is human society now so good that any ethical man could wish it to persist unchanged or could fail to hope and to work for its improvement. On the biological side, few inhabitors of a human body can possibly think that it is perfect and that some change in it would not be highly desirable. Even those—movie stars, perhaps—who may think their own bodies incapable of improvement at the moment must deplore the approaching ravages of age and must, by the very contrast between themselves and others, perceive that the physical average for mankind needs changing. Whatever one may think of a possible future utopia in which man would be socially and biologically so perfect that any change would be for the worse, further human evolution now is obviously desirable provided, of course, that it is desirable in direction.

Man has the power to modify and within certain rather rigid limits to determine the direction of his own evolution. This power is increasing rapidly as knowledge of evolution increases. As regards biological evolution, this power has not as yet really been exercised systematically and consciously to any effective extent. Control of social evolution has also been much less in the past than it can be and is likely to be in the future, and has likewise been highly unsystematic and often not really conscious.

Invention by man of the new evolution, based on the inheritance of learning and worked out in social structures, has not eliminated in him the old organic evolution. The new evolution continues to interact with and in considerable measure to depend on the old. Guidance of the course of either will inevitably influence the other also, and can be most effective only by coordinated guidance of the two. This involves possible, although I think not necessary, conflict. Guidance of social evolution, to the extent that it is undertaken at all, necessarily follows some adopted ethical standards. That guidance becomes more difficult and its results slower if the ethical standards to be followed, like those discussed in the last chapter, hold that it is bad to impose regimentation or unnecessary compulsion on individuals.

The quickest and most effective guidance of biological evolution, too, could be achieved only by compulsion and therefore must here be held to be ethically bad. It is, nevertheless, possible for slower and less rigid but still, in the long run, effective guidance of both to be achieved in ways ethically good. It is each individual's responsibility to choose what he considers right directions for social and for biological evolution. With increase in knowledge and in its dissemination, there should eventually result sufficient unanimity on these points so that effective evolutionary motion would occur by voluntary individual actions. It is one of the known facts of organic evolution that a very minute incidence of natural selection will, under suitable circumstances, ultimately determine the direction of evolution. This involves the possibility that human evolution could be guided by united action of a small minority, although special conditions would be required for their action to become effective. It would, for instance, be ineffective if opposed by a different trend in the rest of the population, whether self-controlled or under the influence of natural selection only. It would be decisive if the direction of evolution were wholly random in the rest of the population or if it were about equally divided among different trends one of which coincided with that desired by the minority.

Under our ethics, the possibility of man's influencing the direction of his own evolution also involves his responsibility for doing so and for making that direction the best possible. Those ethics themselves define the best direction for social evolution. They also define in large measure the desirable directions for organic evolution, although the definition is indirect and less obvious.

The present organic structure of the human species is obviously consonant with far greater progress in the new evolution toward an ethical social ideal than has yet been achieved. Approach toward such an ideal must, however, also involve some physical evolutionary changes. Men differ greatly in intelligence, in temperament, and in various abilities that are in part, at least, hereditary. In some respects these differences are socially and ethically desirable. Differences in temperament are certainly desirable except for the relatively few temperaments that are incompatible with socially normal living. The good society would involve opportunities for the fullest development of every ethically acceptable variety of temperament. Some differences in degree of intelligence may be desirable, but the point is debatable and in any case the raising of the average nearer to the present maximum is evidently desirable. Such changes as these would require provision of optimum environmental conditions, but could not be achieved by this means alone; genetic selection in the existing variability of the species would also be required. There is ample evidence that intelligence is if not strictly determined at least strictly limited as to potentialities by inheritance. Even temperament is apparently to some degree hereditary and correlated with physical structure, better knowledge of which would facilitate selection.[2]

Such changes, involving differences in distribution of existing characters within the human population, are evolutionary but they are of limited scope. Ultimate progress beyond these limits would necessarily involve the development of new characters in the human organism. Probably the new character most surely necessary for evolution beyond the

2. The relative roles of heredity and environment on both physical and temperamental characteristics and the correlation between the two are particularly well discussed in Dobzhansky's *Mankind Evolving*, previously cited. On intelligence, see also J. McV. Hunt, *Intelligence and Experience* (New York, Ronald Press, 1961).

present limits is an increase in intelligence above the existing maximum. Human progress depends on knowledge and learning, and the capacity for these is conditioned by intelligence. Most scientists are already aware that the progress of science is being impeded by the fact that the most brilliant men simply do not have enough learning capacity to acquire all the details of more than increasingly narrow segments of the field of knowledge. Only a very stupid person can believe that mankind is already intelligent enough for its own good.

If and as it is achieved, this increase in intelligence will have other concomitants, among them a larger brain and changes to accommodate this, perhaps in the direction of making the adult more childlike in proportions.[3] Increase in length of life, besides corresponding with a nearly universal human desire, will also ultimately be necessary for intellectual and social progress, in order to allow for a longer juvenile learning period without subtracting from the period of adult life.

Many progressive changes which in other organisms would have to be physical, do not seem necessary or even desirable in man because they correspond with needs that he can supply more effectively and rapidly by technological means. There is no possible substitute for intelligence, but it can be greatly aided and supplemented. Organic development of new perceptual or, at least, sensory apparatus does not seem

3. It is well known that man today resembles the young more than the adults of other higher primates. On this and other points pertinent to the present chapter, see Haldane's provocative paper, "Man's Evolution: Past and Future" (*Atlantic Monthly, 179* [1947], 45–51). The paper was originally delivered at a conference the whole of which was of interest in relationship to the present enquiry and which was published as G. L. Jepsen, E. Mayr, and G. G. Simpson, editors, *Genetics, Paleontology and Evolution* (Princeton, Princeton Univ. Press, 1949). I have also discussed the possible biological future of man in Chapter 14 of *This View of Life,* previously cited.

to be required for future human evolution. I am personally skeptical as to whether extrasensory perception, the other so-called psi phenomena, "group consciousness," and other such things dear to the hearts of many writers on the present and future of man really exist. Even if they do there is no reason to develop them because, as far as they might have any use, the same results can be achieved more fully and surely by mechanical means. It is, indeed, quite possible that development of such supposed phenomena would impede ethical progress.

The means of achieving biological evolutionary progress are already becoming clear, although it is doubtful whether we are yet ready to apply them well. The known avenues of such change are by environmental conditions of development, by selection, and by mutation (followed, of course, by selection). Control of environmental conditions is the only means now commonly in use and it is definitely advancing, although much remains to be done. It permits exploration of the full potentialities of the human organism in the way of healthier bodies, realization of more nearly the full life span, and so on, but it does not change those potentialities.[4]

Further steps, if and when taken, must involve selection, that is, some degree of control over differential reproduction. In principle this could be completely controlled by man, but even partially effective control is almost impossible in the present state of society and it is doubtful whether really full control could ever be exercised in an ethically good social system. The right to apply it must be voluntarily granted by the individuals concerned. Its effectiveness and the determination of the right direction in which to apply it will demand a great deal of knowledge that we do not now

4. "Control of environmental conditions" of course includes prevention and cure of internal diseases, proper diet, and other factors of health, as well as full and proper development of the mind and of social relationships.

possess. Eugenics has deservedly been given a bad name by many sober students in recent years because of the prematurity of some eugenical claims and the stupidity of some of the postulates and enthusiasms of what had nearly become a cult. We are also still far too familiar with some of the supposedly eugenical practices of the Nazis and their like. The assumption that biological superiority is correlated with color of skin, with religious belief, with social status, or with success in business is imbecile in theory and vicious in practice. The almost equally naïve, but less stupid and not especially vicious, idea that prevention of reproduction among persons with particular undesirable traits would quickly eradicate these traits in the population has also proved to be unfounded. The incidence of a few clearly harmful hereditary defects could be reduced by sterilization of the individuals possessing them, but they could not be wholly eliminated and in the light of present knowledge it is highly doubtful whether this means can produce any really noteworthy physical improvement in the human species as a whole.

Selection was, nevertheless, the means by which man arose and it is the means by which, if by any, his further organic evolution must be controlled.[5] Man has so largely modified the impact of the sort of natural selection which produced him that desirable biological progression on this basis is not to be expected. There is no reason to believe that individuals with more desirable genetic characteristics now have more children than do those whose genetic factors are undesirable, and there is some reason to suspect the opposite.

5. Irresponsible extrapolators from current knowledge of molecular genetics have said that individual human genetic systems may soon be constructed to order by artificial synthesis of genes. In fact it is not yet clear whether that is theoretically possible, and it certainly is impracticable by any present techniques or in the even distantly foreseeable future. See T. M. Sonneborn, "Implications of the New Genetics for Biology and Man" (*A.I.B.S. Bulletin, 13* [1963], 22–26).

The present influence of natural selection on man is at least as likely to be retrogressive as progressive. Maintenance of something near the present biological level is probably about the best to be hoped for on this basis. The only proper possibility of progress seems to be in voluntary, positive social selection to produce in offspring new and improved genetic systems and to balance differential reproduction in favor of those having desirable genes and systems. As soon as we know what the desirable human genes and systems are and how to recognize them! The knowledge is now almost wholly lacking, but it seems practically certain that it is obtainable.

One thing that is definitely known now is that breeding for uniformity of type and for elimination of variability in the human species would be ethically, socially, and genetically bad and would not promote desirable evolution. This variability, with accompanying flexibility and capacity for individualization, is in itself ethically good, socially valuable, and evolutionarily desirable. It happens that the present human breeding structure is excellent for the promotion of adaptability and desirable variability and for control of evolution by selection. The theoretically ideal conditions in this respect involve a large population with wide genetic variety (reflected also in local polymorphism), divided into many relatively small, habitually interbreeding groups which are not, however, completely isolated but also have some gene interchange between them. This ideal is actually rather closely approached in the present breeding structure of the human species. Its continuance as a basis for effective selection and maintenance of desirable variability demands avoidance of both of two extremes. On one hand, a completely classless society or habitual general intermingling in marriage of all racial or other groups would be bad, from this point of view. On the other hand, effective segregation and prohibition of interbreeding between any two or more racial, religious, or other groups would be even worse.

Selection can in the long run make the most of the genetic factors now existing in the human species. Eventually its possibilities would be exhausted if there were not also new mutations. Mutations are known to occur in mankind at rates comparable to those in other animals and consistent with sustained evolution at moderate speed. Almost all of those known are disadvantageous and produce abnormalities definitely undesirable in present society and probably of no value for any desirable future development.[6] We probably completely miss in study of human heredity the very small favorable mutations that are likely to exist and to be more frequent than these larger and generally unfavorable mutations. The ability to recognize these small mutations will be a necessary factor if man is to advance his own biological evolution by voluntary selection. At present we do not have the slightest idea as to how to produce to order the sort of mutation that may be needed or desired. We do not even know whether this is physically possible. If it is, and if man does discover the secret, then indeed evolution will pass fully into his control.

Now we cannot predict for sure whether the future course of human evolution will be upward or downward. We have, however, established the fact that it *can* be upward and we have a glimpse, although very far from full understanding, as to how to ensure this. It is our responsibility and that of our descendants to ensure that the future of the species is progressive and not retrogressive. The immediate tasks are to work for continuance of our species, for avoiding early self-extinction, settling ideological battles, and progressing toward an ethically good world state. The immediate means

6. Therefore on the whole the currently accelerating induction of mutations by artificially increased exposure to mutagenic radiation augurs ill for mankind. Opinions differ extremely and hotly on the seriousness of that danger. For a sensibly balanced opinion see B. Wallace and Th. Dobzhansky, *Radiation, Genes, and Man* (New York, Henry Holt, 1959).

not only for these tremendous tasks but also for the future task of guiding human evolution lie within the ethic of knowledge. We need desperately to know more about ourselves, about our societies, about all of life, about the earth, and about the universe. We need to balance our knowledge better, to reverse the disparity in discovery in the physical, biological, and social sciences so that the social sciences shall be first and the physical last. We need to realize more fully and widely that technological advances and the invention and enjoyment of gadgets are not the most useful sort of knowledge and are relatively quite unimportant (occasionally downright harmful) for true human progress. We need to remember that cultural evolution proceeds only by interthinking, as organic evolution does only by interbreeding. The most brilliant of geniuses is an intellectual eunuch if his knowledge is not disseminated as widely as possible. It is immoral for any man, industry, or nation to reserve knowledge for its own advantage alone.

We need, too, to recognize the supreme importance of knowledge of organic and of social evolution. Such knowledge provides most of what we know of our place in the universe and it must guide us if we are to control the future evolution of mankind.

EPILOGUE AND SUMMARY

It has been a long journey down the corridors of time to this point where mankind looks with foreboding and yet with hope into the mists of the future. We have followed three billion years of the history of life; we have examined some of the processes by which life has changed and developed; and we have come at the end to consideration of ourselves, the only form of life that knows it has a history, the thinking, responsible, and ethical animals.

Among the questions that have been discussed are some that everyone has asked, in one form or another, and for which everyone has had to find some sort of answer, whether vague or definite, unthinkingly accepted or carefully formulated. This is true because of the very fact that we are human beings and that it is a characteristic of humanity to be conscious of self, to have some sort of concept of what man is and some sort of judgment of his actions. There are many possible approaches to the problems of existence and of values. All should be followed, for none can hope to achieve a total solution and each may be expected to illumine a different aspect of the riddle of human life. Here we have started with the premise that life has evolved and has had a history. This premise has been so conclusively established by generations of study and the resultant accumulation of literally millions of concordant facts that it has become almost self-evident and requires no further proof to anyone reasonably free of old illusions and prejudices.

This premise taken as established, an enquiry into the nature and meaning of life based on it seems, certainly not the only one possible or required, but the most direct and the most fundamental as basis for any further enquiry. This nature and this meaning must, if anywhere, be apparent in the history that has brought man and all other forms of life into being and in the processes of that history. The general outline of that history and some of its characteristic details are now so well determined as to provide a factual background open to little serious question. It is, however, still true that the unknown exceeds the known and gives room for some (yet for limited) differences of interpretation. And even were all factually known, which can never become true, interpretation would still be necessary before meaning could arise from the factual record. Differences of interpretation will no doubt always arise, and this or any other reading of meaning into the history of life can never carry compulsive authority. It can only be an opinion submitted for judgment and each of us is still required to exercise his responsibility to take nothing on faith but to make rational choice.

The existence of this personal responsibility for choice and decision is one of the conclusions of the present study, and it applies to the study itself and to the readers of it. Yet it would be improper to present this with undue humility. Any conclusions, certainly including those here reached, may be false as well as true and any may be responsibly rejected as well as accepted by those to whom they are presented. If, however, evidence is given for those conclusions and the ways of interpreting that evidence are set forth, no one can honestly turn to the conclusions and say, "This I will believe," or "That I refuse to believe," without some weighing of reasons and some positive knowledge to confirm his decisions.[1]

1. Human nature being what it is, there will surely be some who turn first to these concluding remarks in the hope of learning what this book may have to offer without plowing through the preceding discussion. It would be the most humorless arrogance to attempt to order them

What has been said on previous pages is already a summary, and an exceedingly condensed summary. The subject has such tremendous scope and complexity that it would be hard to name another comparable to it in these respects. Summarizing requires simplification and with such a subject any simplification runs the risk of being oversimplification. A summary of the summary cannot possibly avoid this danger, but comprehension requires a guide to main themes, even though the complexities must remain understood rather than expressed, and this is the more necessary the greater those complexities are. Such guidance to some of the main themes should, then, be attempted.

The history of life is to be studied by a great variety of means, among which special importance attaches to the actual historical record in rocks and the fossils contained in them. The rocks indicate that the earth, only home of life known to us, has existed in approximately its present form and with roughly the same physical surface conditions for more than 3,000,000,000 years. Direct record of life extends back to about that time. The very beginnings of life are not known in the record and in the nature of things could not be adequately recorded, if at all. Studies on other bases demonstrate not only the possibility but also the probability that life arose from the inorganic spontaneously, that is, without supernatural intervention and by the operation of material processes, themselves of unknown origin, sometime during the first billion years or so of the earth's existence. After long obscurity during which the basic complexity of cellular organization was developed, there came a time, extending over at least several tens of millions of years, during which fundamental divergence produced most of the broadest structural

---

to turn back to page one, but they should be told that one of the conclusions is that there is no ethical right to judge without consideration of the basis for judgment.

and functional types of living organisms that were ever to exist. This occurred around 600,000,000 years ago, and since that time life has left a continuous and fairly full record of its own history.

Few, if any, of the broadest and most basic types have ever become extinct. The more complex and, in some sense, higher types have in some cases changed greatly and on the whole have changed more than the lower types. All main types represent abilities to follow broadly distinctive ways of life, and the earlier or lower persist along with the later and higher because these latter represent not competitors doing the same sorts of things as their lower ancestors but groups developing distinctly new ways of life. Each main group shows, in the record, at least one period of major expansion when it was (or is) particularly abundant and varied. Expansion and contraction in different groups follow highly diverse ways and there is no suggestion of a standard pattern or over-all plan. The sum of all these effects, the total number and variety of organisms existing in the world, has shown a tendency to increase markedly during the history of life, but even this has shown much inconstancy and fluctuation.

Expansions of a given group commonly occur by radiation into a variety of related ways of life. When groups living under quite different conditions and with unrelated ways of life expand in this way, there is no evident tendency for their expansions either to coincide or to be successive. Within any one general way of life there is a tendency for successive expansions, each to some extent replacing the last but often not completely eradicating it. Among land vertebrates, with broadly related ways of life, there have been successive expansions of amphibians, reptiles, and mammals. Primates arose in the main mammalian radiation and had themselves, within the group, a succession of radiations partly replacing older groups and partly extending into new ways of life. Man arose from the most recent of these primate radiations.

In attempting to interpret this history the major problem, both philosophical and scientific, is to decide whether it has taken place under the action of universal and natural principles, and so is naturalistic; whether it has involved principles new in and peculiar to life, making it vitalistic; and whether, in either case, it does or does not represent the working out of some supernal purpose, involving an over-all plan and progressing toward a goal, the finalistic interpretation.

Examined in more detail, the history of life turns out to be an odd and intricate mixture of the oriented and the random. Continuing and clearly oriented trends of evolutionary change are very common, but when carefully studied without gross oversimplification, they give no appearance of rigid control forcing them in only one direction. They also lack evidence of any vital inner force or momentum that carries them forward regardless of the functional adaptation to way of life or of any random change. The evidence all concurs in suggesting that the orienting force in evolution is neither internal nor external to the organisms involved but is in that interplay of both internal and external factors which produces adaptation to way of life and to environment.

Evolution appears to be not only a mixture of random and of oriented changes but also highly opportunistic, in a purely impersonal sense. Most, although not all, of the possible ways of life in any given period of earth history have usually been followed by one group or another. Among possible different solutions of a given functional problem, such as that of perception of light, many or all are commonly followed by various groups. There is no effect of over-all plan tending toward the same solution of such a problem. Solutions are not achieved in the way theoretically best but on the basis of what happens to be available. Patterns of racial life and death, of survival and extinction, expansion and contraction, are also extremely varied and are opportunistic, correlated with the possibilities of existing physical factors, and not with any conceivable over-all plan or ultimate purpose.

Although many details remain to be worked out, it is already evident that all the objective phenomena of the history of life can be explained by purely naturalistic or, in a proper sense of the sometimes abused word, materialistic factors. They are readily explicable on the basis of differential reproduction in populations (the main factor in the modern conception of natural selection) and of the mainly random interplay of the known processes of heredity. Vitalist and finalist theories not only fail to explain them but are also flatly inconsistent with them. These theories are commonly not truly explanatory but are frequently more or less veiled attempts to evade explanation. They arose in part as efforts to salvage unscientific prejudices really contradicted by the facts of evolution, but some were also legitimate reactions to the fact that proposed materialistic explanations were long unsatisfactory and that the effort to complete them was in a temporary impasse. Discovery of the facts of genetics and integration of these with knowledge of life from other fields of study have led out of this impasse and produced a theory that no longer gives motive for vitalistic or finalistic evasions.

All-over progress, and particularly progress toward any goal or fixed point, can no longer be considered as characteristic of evolution or even as inherent in it. Progress does exist in the history of life, but it is of many different sorts and each sort occurs separately in many different lines. One sort of progress in structure and function that stands out as particularly widespread and important is increasing awareness of the life situation of the individual organism and increasing variety and sureness of appropriate reactions to this. Among the many different lines that show progress in this respect, the line leading to man reaches much the highest level yet developed. By many other criteria of progress, also, man is at least among the higher animals and a balance of considerations warrants considering him the highest of all.

Man is the result of a purposeless and natural process that did not have him in mind. He was not planned. He is a state of matter, a form of life, a sort of animal, and a species of the Order Primates, akin nearly or remotely to all of life and indeed to all that is material. It is, however, a gross misrepresentation to say that he is *just* an accident or *nothing but* an animal. Among all the myriad forms of matter and of life on the earth, or as far as we know in the universe, man is unique. He happens to represent the highest form of organization of matter and energy that has ever appeared. Recognition of this kinship with the rest of the universe is necessary for understanding him, but his essential nature is defined by qualities found nowhere else, not by those he has in common with apes, fishes, trees, fire, or anything other than himself.

It is part of this unique status that in man a new form of evolution begins, overlying and largely dominating the old, organic evolution which nevertheless also continues in him. This new form of evolution works in the social structure, as the old evolution does in the breeding population structure, and it depends on learning, the inheritance of knowledge, as the old does on physical inheritance. Its possibility arises from man's intelligence and associated flexibility of response. His reactions depend far less than other organisms' on physically inherited factors, far more on learning and on perception of immediate and of new situations. His unique status also involves an unprecedented degree of foresight and the development of a symbolic system of communication, language, entirely absent in any other now living animals.

Intelligence and foresight bring with them the power and the need for constant choice between different courses of action. Man plans and has purposes. Plan, purpose, goal, all absent in evolution to this point, enter with the coming of man and are inherent in the new evolution, which is confined to him. With them and with foresight comes the need for criteria of choice. Good and evil, right and wrong, concepts irrelevant in nature except from the human viewpoint, become real and pressing features of the whole cosmos as viewed by man—the only possible way in which the cosmos can be viewed morally because morals arise only in man.

Discovery that the universe apart from man or before his coming lacks and lacked any purpose or plan has the inevitable corollary that the workings of the universe cannot provide any automatic, universal, eternal, or absolute ethical criteria of right and wrong. This discovery has completely undermined all older attempts to find an intuitive ethic or to accept such an ethic as revelation. It has not been so generally recognized that it equally undermines attempts to find a naturalistic ethic which will flow with absolute validity from the workings of Nature or of Evolution as a new Revelation. Such attempts, arising from discovery of the baselessness of intuitive ethics, have commonly fallen into the same mistake of seeking an absolute ethic or one outside of man's own nature and have then been doomed to failure by their own premises.

I was at this point in the first draft of this book, when I coincidentally came across some highly pertinent remarks in an example of the legions of articles deploring the decline of religious faith.[2] The author, a distinguished philosopher, found himself agreeing with certain ecclesiastical dignitaries that chaos and bewilderment in the world today result from loss of faith in God and religion. This has become almost banal by constant repetition (although I beg leave to note that repetition does not establish truth). From this point,

2. W. T. Stace, "Man against Darkness," *Atlantic Monthly, 182,* No. 3 (1948), 53–58.

however, the author took a less beaten path and one not likely to comfort the godly. He found that, indeed, the old religious faith was unjustified and that the truth is quite otherwise. He did not question or even particularly deplore the fact that the universe does not operate by divine plan, but he thought it a great pity that we ever found this out. He was a little petulant with scientists for discovering that the world is purposeless and for thus forcing abandonment of religions that require the postulate of purpose. He could only face the fact that childish dreams of a meaningful universe must be laid aside, and he exhorted mankind to become adult and to live as honorably as may be in a stark and bleak world.

The honesty of this philosophical acceptance of truth as we see it must be admired, but it does seem that the total surrender was premature and that his own argument was still not entirely adult, in that author's sense of the word. It seems still a little childish to regret that vain dreams have ended. Perhaps it is even more juvenile to blame the loss rather on the scientists (unkind adults) who exposed the sham than on the falseness of the dreams or on the dreamers. Most important of all, the fully adult reaction to the loss of false ideals would seem to be not simply to regret them and to determine grimly to face the world without them, but to see whether true values may not exist and whether they may not indeed be easier to come by now that the false are cleared away. It is the final main theme of the present enquiry that such values do, indeed, exist.

The ethical need is within and peculiar to man, and its fulfillment also lies in man's nature, relative to him and to his evolution, not external or unchanging. Man has choice and responsibility, and in this matter, too, he must choose and he cannot place responsibility for rightness and wrongness on God or on nature. The choices here proposed and placed before the reader for his acceptance or rejection in accordance with his own responsibility are, first, based on the proposition that knowledge is good and, second, derived from the fact that man possesses personal responsibility. In all, these ethics flow from the unique qualities of man and from what is conceived to be his real place in nature.

The fact of responsibility and the ethic of knowledge have many ethical corollaries, among them that blind faith ("blind," or "unreasoning," to be emphasized) is morally wrong. In connection with high individualization, another human diagnostic character, the resulting ethics include the goodness of maintenance of this individualization and promotion of the integrity and dignity of the individual. Socialization, a necessary human process, may be good or bad. When ethically good, it is based on and in turn gives maximum *total* possibility for ethically good individualization.

Such ethics have wide applications in social and personal conduct. They stand in strong opposition to authoritarian or totalitarian ideologies. They confirm the existence of many evils in current democracies, but the good state, on these principles, would inevitably be a democracy. The principles do not label every human action as good or bad—to do this would violate the prior conclusion that valid ethics cannot be absolute. Moreover, individual variability and flexibility are in themselves desirable, from the point of view both of ethics and of evolution, biological and social.

Beyond the subjects of this enquiry are mysteries deeper still, unplumbed by paleontology or any other science. The boundaries of what can be achieved by perception and by reason are respected. The present chaotic stage of humanity is not, as some wishfully maintain, caused by lack of faith but by too much unreasoning faith and too many conflicting faiths within these boundaries where such faith should have no place. The chaos is one that only responsible human knowledge can reduce to order.

It is another unique quality of man that he, for the first time in the history of life, has increasing power to choose his course and to influence his own future evolution. It would be rash, indeed, to attempt to predict his choice. The possibility of choice can be shown to exist. This makes rational the hope that choice may sometime lead to what is good and right for man. Responsibility for defining and for seeking that end belongs to all of us.

INDEX

Asterisks denote illustrations